Industrial Internet of Things

Artificial Intelligence (AI): Elementary to Advanced Practices

Series Editors: Vijender Kumar Solanki, Zhongyu (Joan) Lu, and Valentina E Balas

In the emerging smart city technology and industries, the role of artificial intelligence is getting more prominent. This AI book series will aim to cover the latest AI work, which will help the naïve user to get support to solve existing problems and for the experienced AI practitioners, it will assist to shedding light for new avenues in the AI domains. The series will cover the recent work carried out in AI and its associated domains, it will cover Logics, Pattern Recognition, NLP, Expert Systems, Machine Learning, Block-Chain, and Big Data. The work domain of AI is quite deep, so it will be covering the latest trends which are evolving with the concepts of AI and it will be helping those new to the field, practitioners, students, as well as researchers to gain some new insights.

Cyber Defense Mechanisms
Security, Privacy, and Challenges
Gautam Kumar, Dinesh Kumar Saini, and Nguyen Ha Huy Cuong

Artificial Intelligence Trends for Data Analytics Using Machine Learning and Deep Learning Approaches
K. Gayathri Devi, Mamata Rath, and Nguyen Thi Dieu Linh

Transforming Management Using Artificial Intelligence Techniques
Vikas Garg and Rashmi Agrawal

AI and Deep Learning in Biometric Security
Trends, Potential, and Challenges
Gaurav Jaswal, Vivek Kanhangad, and Raghavendra Ramachandra

Enabling Technologies for Next Generation Wireless Communications
Edited by Mohammed Usman, Mohd Wajid, and Mohd Dilshad Ansari

Artificial Intelligence (AI)
Recent Trends and Applications
Edited by S. Kanimozhi Suguna, M. Dhivya, and Sara Paiva

Deep Learning for Biomedical Applications
Edited by Utku Kose, Omer Deperlioglu, and D. Jude Hemanth

Cybersecurity
Ambient Technologies, IoT, and Industry 4.0 Implications
Gautam Kumar, Om Prakash Singh, and Hemraj Saini

Industrial Internet of Things
Technologies, Design, and Applications
Edited by Sudan Jha, Usman Tariq, Gyanendra Prasad Joshi, and Vijender Kumar Solanki

For more information on this series, please visit: https://www.routledge.com/Artificial-Intelligence-AI-Elementary-to-Advanced-Practices/book-series/CRCAIEAP

Industrial Internet of Things

Technologies, Design, and Applications

Edited by
Sudan Jha
Usman Tariq
Gyanendra Prasad Joshi
Vijender Kumar Solanki

CRC Press is an imprint of the
Taylor & Francis Group, an **informa** business

First edition published 2022
by CRC Press
6000 Broken Sound Parkway NW, Suite 300, Boca Raton, FL 33487-2742

and by CRC Press
2 Park Square, Milton Park, Abingdon, Oxon, OX14 4RN

© 2022 selection and editorial matter, Sudan Jha, Usman Tariq, Gyanendra Prasad Joshi, and Vijender Kumar Solanki; individual chapters, the contributors

CRC Press is an imprint of Taylor & Francis Group, LLC

Reasonable efforts have been made to publish reliable data and information, but the author and publisher cannot assume responsibility for the validity of all materials or the consequences of their use. The authors and publishers have attempted to trace the copyright holders of all material reproduced in this publication and apologize to copyright holders if permission to publish in this form has not been obtained. If any copyright material has not been acknowledged please write and let us know so we may rectify in any future reprint.

Except as permitted under U.S. Copyright Law, no part of this book may be reprinted, reproduced, transmitted, or utilized in any form by any electronic, mechanical, or other means, now known or hereafter invented, including photocopying, microfilming, and recording, or in any information storage or retrieval system, without written permission from the publishers.

For permission to photocopy or use material electronically from this work, access www.copyright.com or contact the Copyright Clearance Center, Inc. (CCC), 222 Rosewood Drive, Danvers, MA 01923, 978-750-8400. For works that are not available on CCC please contact mpkbookspermissions@tandf.co.uk

Trademark notice: Product or corporate names may be trademarks or registered trademarks and are used only for identification and explanation without intent to infringe.

ISBN: 978-0-367-60777-7 (hbk)
ISBN: 978-0-367-60867-5 (pbk)
ISBN: 978-1-003-10226-7 (ebk)

DOI: 10.1201/9781003102267

Typeset in Times
by SPi Technologies India Pvt Ltd (Straive)

Contents

Foreword vii
Editors...ix
Contributors xi

Chapter 1   Introduction to Industrial Internet of Things (IIoT)........................ 1

                    Hanan Ahmed, A.A. Ramadan, E.H. Elkordy, and Ahmed A. Elngar

Chapter 2   Challenges in Industrial Internet of Things (IIoT)........................... 19

                    Sarbagya Ratna Shakya and Sudan Jha

Chapter 3   IoT Based Automated Healthcare System 41

                    Darpan Anand and Aashish Kumar

Chapter 4   Internet of Things (IoT)-Based Industrial Monitoring System......... 55

                    Syeda Florence Madina, Md. Shahinur Islam,
                    Fakir Mashque Alamgir, and Mohammad Farhan Ferdous

Chapter 5   Internet Working of Vehicles and Relevant Issues
                    in IoT Environment... 87

                    Rajeev Kumar Patial and Deepak Prashar

Chapter 6   Adoption of Industry 4.0 in Lean Manufacturing........................... 107

                    Nishant Jha and Deepak Prashar

Chapter 7   Internet of Things Based Economical Smart Home
                    Automation System 129

                    Pawandeep Kaur and Krishan Arora

Chapter 8   Machine Vision Technology, Deep Learning and IoT
                    in Agricultural Industry.. 143

                    K. Harjeet and Deepak Prashar

Chapter 9 IIoT Edge Network and Spectrum Scarcity Issue............161

Gyanendra Prasad Joshi and Sudan Jha

Chapter 10 Review on Optical Character Recognition Based
Applications of Industrial IoT 175

Apurva Sonavane and Jimmy Singla

Chapter 11 Using Blockchain in Resolving the Challenges
Faced by IIoT.............................189

Nishant Jha and Deepak Prashar

Chapter 12 Internet of Things-Based Arduino Controlled On-Load
Tap Changer Distribution Transformer............217

Krishan Arora

Index............225

Foreword

Internet of Things (IoT) has evolved as an integral part of our life from Smart appliances, Smart Wearable devices, connected cars, and smart healthcare devices to smart energy meters. The IoT has not only comfort the human life but has prospered the surrounding systems with intelligent services. But since IoT is in the initial stage of development, there are plenty of research opportunities available. The Industrial Internet of Things (IIoT) focuses on the use of cyber-physical systems to monitor the physical factory processes and make data-based automated decisions.

This book covers a wide range of topics related to IIoT, including introduction, tools, algorithms, applications, along with security issues and solutions relevant in the field of IIoT. More importantly, the book gives a glimpse of new research areas in IIoT for the persons who are willing to do research in this domain.

The first edition of this book is designed to impart a basic knowledge of Industry 4.0 and IIoT. Other sections in the book cover:

Introduction to Industrial Internet of Things (IIoT) and its Types
Advances of IoT towards IIoT
Role of IIoT in Information Technology
Security Issues in IoT – how IIoT resolves security issues in IoT
Applications and case studies of IIoT
Key enabling technologies of IIoT

Series Editors
Jha/Tariq/Joshi/Solanki

Editors

Dr. Sudan Jha received his Ph.D. in Computer Science and Engineering, in 2015 from India and M.S. in Computer Science in 2006 from Nepal. His major field of study is "cell throughput and carrier enhancement in IPv6". His research interest also includes Quality assessment of IoT devices, Agent based learning systems in smart devices, and Neutrosophic soft systems applications. He is a senior member of IEEE in 2018 and holds brand Ambassador of Bentham Science Series. Currently, he is a Professor in the School of Computational Science at Christ University, Delhi-NCR Campus, India. He has published research articles in SCI/SCIE/ESCI along with 5 books.

Dr. Usman Tariq is an associate professor with the College of Computer Engineering and Sciences at PSAU. He holds a Ph.D. from the Ajou University, South Korea and led the design of a global data infrastructure simulator modeling, to evaluate the impact of competing architectures on the performance, availability, and reliability of the system for Industrial IoT infrastructure. Currently, he is interested in applied cyber security, advanced topics in Internet of Things, and health informatics. His research focus is on the theory of large complex networks, which includes network algorithms, stochastic networks, network information theory, and large-scale statistical inference.

Dr. Gyanendra Prasad Joshi is an Assistant Professor at the Department of Computer Science and Engineering at Sejong University, South Korea. He worked at Case School of Engineering, Case Western Reserve University, Cleveland, OH from March 2018 to February 2019. He received KRF scholarship and ITSP scholarships from Korean government for his MS and PhD studies. He has served on the review process of several international journals including Elsevier's JNCA, Springer's Wireless Networks, Wiley's ETT, KSII's TIIS, EURASIP JWCN, IEEE Communications Letters, China Communication, IET Communications, AIoT Journal, among others. He has successfully organized many conferences including ICTMHC-2016, ICACCI-2015-2020, and ICIDB-2015-2019. He has more than 50 research articles published in books, international journals, and international conferences as a first author. His main research interests include UAV localization, MAC and routing protocols for next-generation wireless networks, wireless sensor networks, cognitive radio networks, RFID system, IoT, smart city, deep learning and digital convergence.

Dr. Vijender Kumar Solanki (Ph.D.) is an Associate Professor in Computer Science & Engineering, CMR Institute of Technology (Autonomous), Hyderabad, TS, India. He has more than 10 years of academic experience in network security, IoT, Big Data, Smart City and IT. He has attended an orientation program at UGC-Academic Staff College, University of Kerala, Thiruvananthapuram, Kerala & Refresher course at Indian Institute of Information Technology, Allahabad, UP, India. He has authored or co-authored more than 50 research articles that are published in various journals, books and conference proceedings. He has edited or co-edited 14 books and Conference Proceedings in the area of soft computing. He is the Book Series Editor of Internet of Everything (IoE): Security and Privacy Paradigm, CRC Press, Taylor & Francis Group, USA; Artificial Intelligence (AI): Elementary to Advanced Practices Series, CRC Press, Taylor & Francis Group, USA; IT, Management & Operations Research Practices, CRC Press, Taylor & Francis Group, USA; Bio-Medical Engineering: Techniques and Applications with Apple Academic Press, USA and Computational Intelligence and Management Science Paradigm, (Focus Series) CRC Press, Taylor & Francis Group, USA. He is Editor-in-Chief in International Journal of Machine Learning and Networked Collaborative Engineering (IJMLNCE); International Journal of Hyperconnectivity and the Internet of Things (IJHIoT), IGI-Global, USA, Co-Editor Ingenieria Solidaria Journa, Associate Editor in International Journal of Information Retrieval Research (IJIRR), IGI-GLOBAL, USA, He has been guest editor with IGI-Global, USA, InderScience, and other publishers.

Contributors

Hanan Ahmed
Demonstrator of Computer Science
High Institute for Computers and
  Management Information System
New Cairo, Egypt,

Fakir Mashque Alamgir
East West University
Dhaka, Bangladesh

Dr. Darpan Anand
Department of CSE
Chandigarh University
Punjab

Krishan Arora
School of Electronics and Electrical
  Engineering, Lovely Professional
  University Phagwara
Punjab, India

E.H. Elkordy
Assistant Professor of Pure
  Mathematics, Mathematics and
  Computer Science Dept., Faculty of
  Science, Beni-Suef University
Beni-Suef, Egypt

Ahmed A. Elngar
Faculty of Computers and Artificial
  Intelligence Beni-Suef University
Beni-Suef, Egypt

Mohammad Farhan Ferdous
Japan-Bangladesh Robotics and
  Advance Technology Research
  Centre (JBRATRC)
Dhaka, Bangladesh

Gyanendra Prasad Joshi
Sejong University, South Korea
Seoul, South Korea

K. Harjeet
Lovely Professional University,
  Phagwara
Punjab, India

Md. Shahinur Islam
East West University
Dhaka, Bangladesh

Nishant Jha
Lovely Professional University,
  Phagwara
Punjab, India

Sudan Jha
Chandigarh University
Punjab, India

Pawandeep Kaur
School of Electronics and Electrical
  Engineering, Lovely Professional
  University Phagwara
Punjab, India

Mr. Aashish Kumar
Department of CSE
Chandigarh University
Punjab

Syeda Florence Madina
East West University
Dhaka, Bangladesh

Rajeev Kumar Patial
School of Electronics and Electrical
  Engineering, Lovely Professional
  University, Phagwara
Punjab, India

Deepak Prashar
School of Computer Science and
Engineering, Lovely Professional
University Phagwara
Punjab, India

A.A. Ramadan
Professor of Pure Mathematics,
Mathematics and Computer Science
Dept., Faculty of Science
Beni-Suef University
Beni-Suef, Egypt

Sarbagya Ratna Shakya
University of Southern Mississippi
Hattiesburg, Mississippi

Jimmy Singla
Lovely Professional University,
Phagwara
Punjab, India

Apurva Sonavane
Lovely Professional University,
Phagwara
Punjab, India

Jha, Sudan
CHRIST University
Delhi-NCR Campus, Delhi, India

1 Introduction to Industrial Internet of Things (IIoT)

Hanan Ahmed

High Institute for Computers and Management Information System, New Cairo, Egypt

A.A. Ramadan and E.H. Elkordy

Beni-Suef University, Beni-Suef, Egypt

Ahmed A. Elngar

Faculty of Computers and Artificial Intelligence Beni-Suef University, Beni-Suef City, Egypt

CONTENTS

1.1	Industrial Internet of Things	1
1.2	IoT, IIoT and Industry 4.0	2
1.3	IIoT Architectures and Frameworks	5
1.4	Challenges of IIoT	9
1.5	Conclusion	15
References		15

1.1 INDUSTRIAL INTERNET OF THINGS

IIoT serves as a modern vision of IoT within the industrial sector by mechanizing smart objects for detecting, collecting, sensing, handling, and communicating the events of the real-time occasion in industrial systems. The main objective of IIoT is to achieve high operational productivity, expanded efficiency, and superior management of industrial resources and forms through item customization, brilliantly checking applications for generation floor shops and machine well-being, and predictive and preventive support of industrial hardware [3]. In [1], authors have defined the IIoT as follows: "Industrial IoT (IIoT) is the network of intelligent and highly connected industrial components that are deployed to achieve high production rate with reduced operational costs through real-time monitoring, efficient management and controlling of industrial processes, assets and operational time".

DOI: 10.1201/9781003102267-1

IoT and IIoT have their particular focuses on concepts and viable application scenarios in spite of the fact that the IIoT is inferred from the IoT. The IoT broadly acknowledged by individuals is basically consumption-oriented and aims to improve people's quality of life. The foremost normal application scenarios of IoT are keen domestic, well-being observing, and indoor localization. The IIoT is production-oriented and points to make strides in mechanical generation productivity. Ordinary application scenarios of IIoT incorporate perceptive coordination, inaccessible maintenance, and brilliantly production lines. The framework systems of diverse IoT application scenarios, by and large, have to be built from scratch, and the sending scale of sensors is generally little with low exactness requirements [4]. In any case, the framework systems of IIoT application scenarios are built based on the conventional mechanical framework, so the sending scale of sensors is exceptionally expansive with tall accuracy prerequisites. For the IoT, gadgets, for the most part, have solid portability, create medium information volume, and have a tall resilience for the delay; whereas for the IIoT, most of the gadgets are settled in position, produce an incredible sum of seen information, and have a too resistance for the delay [5]. The concept of IIoT is closely related to a few broadly acknowledged concepts, such as cyber-physical frameworks (CPSs), IoT, the Mechanical Web, and Industry 4.0. CPS, proposed by Helen Gill in 2006, emphasizes the profound integration of different data innovations, such as detecting innovation, implanted innovation, and program & equipment innovation, pointing to attain profoundly synergistic and independent informationization capabilities, real-time and adaptable criticism, and positive cycle between the physical and data universes. As a subset of CPS, IoT basically emphasizes the intuition between objects through the Internet based on one of a kind recognizable pieces of proof. The globalization, openness, interoperability, and socialization of the web give the premise for supporting the IoT concept [6]. The Industrial Internet, proposed by the Industrial Internet Consortium (IIC) propelled by the top five companies within the US, to be specific GE, AT&T, IBM, Intel, and Cisco, primarily centers on the development, application, and standardization of imaginative systems, the upgrade of information circulation, and the digital change of the full mechanical field. The subconcept of IIoT called Industry 4.0 propelled in Germany may be an all-inclusive arranged, manufactured intelligence-based data CPS, basically within the shrewd fabricating field. In rundown, CPS gives an outline for the relationship between the physical world and data world based on the interconnection of things, so CPS speaks to the broadest of the concepts specified over. IoT highlights the interconnects among objects through physical addresses, in any case of whether they are industry or civilian situated, whereas the Mechanical Web depicts the potential future patterns of businesses based on rising technologies [6, 7].

1.2 IoT, IIoT AND INDUSTRY 4.0

IoT, IIoT, and Industry 4.0 are closely related concepts but cannot be traded utilized. In this area, we offer a harsh classification of these terms. Concerning the IoT, a few definitions exist, each one attempting to capture one of its essential characteristics. It is frequently considered as a sort of web for the machines, highlighting the point of permitting things to trade information. Nonetheless, application areas so

Introduction to Industrial Internet of Things (IIoT)

different from a few necessities (particularly those related to communication angles) can be exceptionally diverse, depending on the planning objectives and end-users, the basic commerce models, and the received mechanical arrangements [7]. What is usually addressed as IoT could be better named as consumer IoT, as opposed to IIoT. Customer IoT is human-centered; the "things" are savvy buyer electronic gadgets interconnected with each other in arrange to make strides human mindfulness of the encompassing environment, sparing time and cash. In common, customer IoT communications can be classified as machine-to-user and within the shape of client–server intelligence. On the other hand, within the industrial world, we are helping to the approach of the computerized and shrewd fabricating, which point at coordination Operational Innovation (OT) with Data Innovation (IT) spaces [8]. In exceptionally few words, the IIoT (the fundamental column of advanced fabricating) is almost interfacing all the mechanical resources, counting machines, and control frameworks, with the data frameworks and the commerce forms [8]. As a result, the huge amount of information collected can nourish analytics arrangements and lead to ideal mechanical operations. On the other hand, shrewd fabricating clearly centers on the fabricating arrange of (shrewd) items life cycle, with the objective of rapidly and powerfully react to request changes. Subsequently, the IIoT affects all the mechanical esteem chain and could be a necessity for savvy manufacturing. As underlined within the taking after, communication in IIoT is machine arranged and can extend over a huge assortment of distinctive showcase divisions and exercises. The IIoT scenarios incorporate bequest observing applications (e.g., handle checking in generation plants) and imaginative approaches for self-organizing frameworks (e.g., autonomic mechanical plant that requires small, in case any, human intervention) [9]. Whereas the foremost common communication prerequisites of IoT and IIoT are comparable, e.g., bolster for the web biological system utilizing low-cost, resource-constrained gadgets and organize versatility, numerous communication necessities are particular to each space and can be exceptionally distinctive, e.g., Quality of Benefit (QoS) (in terms of determinism, inactivity, throughput, etc.), the accessibility and unwavering quality, and the security and privacy [8, 9]. IoT centers more on the plan of modern communication benchmarks, which can interface novel gadgets into the web environment in an adaptable and user-friendly way. By differentiating, the current plan of IIoT emphasizes conceivable integration and interconnection of once disconnected plants and working islands or indeed machinery, thus advertising a more proficient generation and unused administrations [9]. For this reason, compared with IoT, IIoT can be considered more of an advancement instead of an insurgency. Table I gives a subjective comparison of these innovations. Regarding the network and criticality, IoT is more adaptable, permitting *ad hoc* and portable arrange structures and having less exacting timing and unwavering quality necessities (but for therapeutic applications) [10]. On the other hand, IIoT regularly utilizes settled and infrastructure-based organization that are well planned to coordinate communication and coexistence needs. In IIoT, communications are within the frame of machine-to-machine joins that got to fulfill rigid prerequisites in terms of convenience and unwavering quality. Taking prepare robotization as a case space where handle observing and control applications can be assembled into three subcategories: monitoring/supervision, closed circle control, and interlocking and control [11].

Whereas observing and supervision applications are less touchy to parcel misfortune and jitter and can endure transmission delay at moment level, closed circle control and interlocking and control applications require bounded delay at the millisecond level (10–100 ms) and a transmission unwavering quality of 99.99%. Comparing the information volume, the created information from IoT is intensely application subordinate, whereas IIoT right now targets analytics. The concept of Industry 4.0 (where 4.0 speaks to the fourth industrial revolution) emerges when the IoT worldview is blended with the CPSs thought [10]. Initially characterized in Germany, the Industry 4.0 concept has picked up a worldwide permeability and it is these days all around embraced for tending to the utilize of web innovations to move forward generation proficiency by implies of shrewd administrations in shrewd industrial facilities. CPSs amplify real-world, physical objects by interconnection them through and through and giving their computerized depictions. Such data, put away in models and information objects that can be upgraded in genuine time, speak to a moment personality of the question itself and constitutes a sort of "digital twin". Thanks to the dynamic nature of these advanced twins, inventive administrations that were not conceivable in the past can be executed over the total item lifecycle, from inception to transfer of fabricated items [11]. IIoT could be a subset of IoT, which is particular to mechanical applications. The fabricating stage of the item lifecycle is where the IoT and Industry 4.0 meet, starting with the IIoT. Figure 1.1 shows convergences of IoT, CPS, IIoT, and Industry 4.0. It must be highlighted that the IIoT worldview isn't intended for substituting conventional computerization applications, but points at expanding the information around the physical framework of intrigued.

As a result, the IIoT (at slightest nowadays) isn't related to controlling applications at the field level, where bounded response time (i.e. determinism) must be guaranteed. On the opposite, as already expressed, IIoT applications counting supervision, optimization, and expectation exercises are ordinarily assembled into the so-called computerized or cloud fabricating (CM). The development intrigued by this subject is affirmed by the wide extend of writing. A study around CM is detailed in [10]. Within the past, the supervision exercises were ruled by man, but proficient machine-to-machine communications make human intercession pointless and expand the working run to a geological scale. For occasion, the accessibility of solid, brief

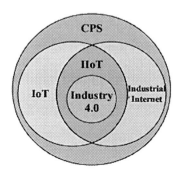

FIGURE 1.1 CPS, IoT, Industrial Internet, IIoT and Industry 4.0 in venn diagram. (Based on Tie Qiu, 2020.)

Introduction to Industrial Internet of Things (IIoT)

idleness associations on such an expansive scale may increment the income [11]. The work in [12] highlights the significance of a real-time, large-scale approach for hardware upkeep applications. An IIoT-based energetic generation coordination engineering is displayed in [12] for real-time synchronization of inside and open generation coordination assets. In [13], the optimization of generation planning is based on IIoT decentralized vitality forecast calculations encouraged by the current state of the machines. Finally, the dynamic decrease of idleness and jitter of the Internet-based network will increment the extent of conceivable applications, as detailed in [14].

1.3 IIoT ARCHITECTURES AND FRAMEWORKS

A nonexclusive architecture of IIoT frameworks was talked about by the industrial web consortium [15] which is displayed in Figure 1.2 where IIoT gadgets and mechanical information sources create persistent information streams at Layer 1, whereas the edge servers and cloud computing frameworks enable IIoT applications at Layer 2 and Layer 3, separately. The endeavor applications are portrayed at Layer 4. Figure 1.2 appears the stream of information and data among diverse layers as well because it shows the organization stream for asset administration and operational flow for overseeing resources within the mechanical systems. In any case, diverse analysts unexpectedly see these designs considering plan varieties in terms of area mindfulness, communication standards, computational assignments, execution ideal models, asset administration plans, security, security, protection, accessibility, and versatility, to title some. Campobello et al. [16] have proposed an arrangement for IIoT named Remote Advancement for Computerization (WEVA) that's based on open-source computer programs and communication conventions. Its design comprises sensors, actuator sheets, bits and working framework, conventions, get to portal, administrations, and applications. Besides, WEVA employs simple WSN as a graphical administration apparatus. The creators suggest that IPv6 may be a prerequisite for IIoT in terms of adaptability. In any case, consolidating these network advances isn't a simple work in arrange to realize a high-performance IIoT in terms of inactivity, security, etc. Numerous analysts have proposed arrangements; nonetheless, they address a particular execution issue and overlook the integration of Remote Sensors Organize (WSN) which plays an imperative part in mechanical applications. Lee et al. [17] have proposed an IIoT suite to attain re-industrialization for Hong Kong by tending to different challenges like objects distinguishing proof in real-time and their areas all through the fabricating forms build up an organized framework that permits objects to communicate between the arrange and other objects in genuine time. Most components of the IIoT suite design incorporate a shrewd center and a cloud stage. The shrewd center works as a door for IoT gadgets and oversees IoT devices in diverse areas. The keen center is serious to achieve three assignments. To begin with, it encourages communication, information trade, and information handling between IoT gadgets. Further, it also gives helpful arrangements when scaling the framework with unused IoT gadgets. At last, it gives a secure association channel between IoT gadgets and the cloud stage by performing information collection, sifting, hostility, and organizing. The cloud stage of IIoT acts just like the brain of the

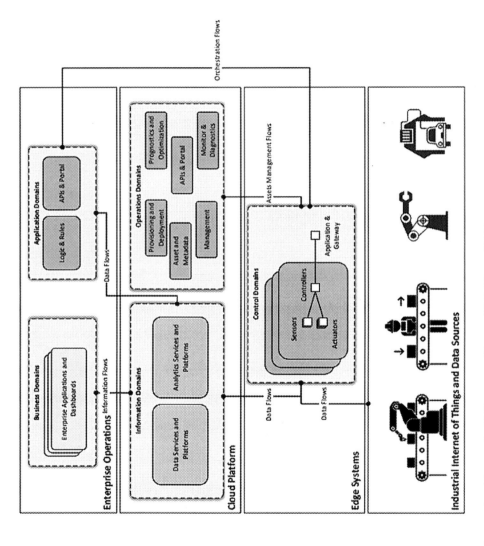

FIGURE 1.2 A general architecture for IIoT systems. (Source: W. Z. Khana, 2019.)

IIoT suite, and is dependable for performing Distinguishing proof and Get to Administration (IAM), stack adjusting, gadget discovery/configuration, directing calculation, checking, and controlling IoT gadgets. Khan et al. [18] have proposed an IoT-based design for controlling and checking oil and gas industry operations. The proposed engineering can be connected to the operations of all three segments (i.e. downstream, midstream, and upstream) of the oil and gas industry. The design comprises three modules counting a shrewd question, portal, and control center. Each module performs extraordinary usefulness and comprises three layers that incorporate an application layer, and arrange layer, and a detecting layer. Keen objects are introduced on diverse oil and gas hardware (e.g., pipelines, capacity tanks, pumps, and wellheads). These smart objects are prepared with diverse sorts of sensors (stream, weight, temperature, and acoustic) to distinguish distinctive occasions like spills, fire, and liquid level. Savvy objects send their detected information specifically or through a portal to the control center. Savvy objects and portals are moreover prepared with radio handset (brief and long extend). The control center comprises databases for information capacity, administration applications, savvy protest interfacing, information examination, and information visualization tools. The arrangement, setup, and interaction between heterogeneous IIoT gadgets are critical issues. To manage these issues, Tao et al. [19] have proposed an IIoT-based center called the hub. The IIHub comprises three modules. IoT centers more on the plan of modern communication benchmarks which can interface novel gadgets into the web environment in an adaptable and user-friendly way. By differentiating, the current plan of IIoT emphasizes conceivable integration and interconnection of once disconnected plants and working islands or indeed machinery, thus advertising a more proficient generation and unused administrations [9]. For this reason, compared with IoT, IIoT can be considered more an advancement instead of an insurgency. Table I gives a subjective comparison of these innovations. Regarding the network and criticality, IoT is more adaptable, permitting *ad hoc* and portable arrange structures and having less exacting timing and unwavering quality necessities (but for therapeutic applications) [10]. On the other hand, IIoT regularly utilizes settled and infrastructure-based organize arrangements that are well planned to coordinate communication and coexistence needs. In IIoT, communications are within the frame of machine-to-machine joins that got to fulfill rigid prerequisites in terms of convenience and unwavering quality. Taking prepare robotization as a case space where handle observing and control applications can be assembled into three subcategories: monitoring/supervision, closed circle control, and interlocking and control [11]. Whereas observing and supervision applications are less touchy to parcel misfortune and jitter and can endure transmission delay at moment level, closed circle control and interlocking and control applications require bounded delay at the millisecond level (10–100 ms) and a transmission unwavering quality of 99.99%. Comparing the information volume, the created information from IoT is intensely application subordinate, whereas IIoT right now targets analytics. The concept of Industry 4.0 (where 4.0 speaks to the fourth industrial revolution) emerges when the IoT worldview is blended with the CPS thought [10]. Initially characterized in Germany, the Industry 4.0 concept has picked up a worldwide permeability and it is these days all around embraced for tending to the utilize of web innovations to move forward generation proficiency by implies of shrewd administrations in shrewd industrial facilities. CPSs amplify real-world, physical objects by interconnection them through and through and giving their computerized depictions. Such data, put away in models and information objects that can be upgraded in genuine time, speak to a moment personality of the question itself and constitutes a sort of "digital twin". Thanks to the dynamic nature of these advanced twins, inventive administrations, that were not conceivable within the past, can be executed over the total item lifecycle, from inception to transfer of fabricated items [11]. IIoT could be a subset of IoT which is particular to mechanical applications. The fabricating stage of the item lifecycle is where the IoT and Industry 4.0 meet, starting with the IIoT. Figure 1.1 appears convergences of IoT, CPS, IIoT, and Industry 4.0. It must be highlighted that the IIoT worldview isn't intended for substituting conventional computerization applications, but points at expanding the information around the physical framework of intrigued. As a result, the IIoT (at slightest nowadays) isn't related to controlling applications at the field level, where bounded response time (i.e. determinism) must be guaranteed. On the opposite, as already expressed, IIoT applications counting supervision, optimization, and expectation exercises, are ordinarily assembled into the so-called Computerized or Cloud Fabricating (CM). The development intrigued by this subject is affirmed by the wide extend of writing. A study around CM is detailed in [10]. Within the past, the supervision exercises were ruled by man, but proficient machine to machine communications make human intercession pointless and expand the working run to a geological scale. For occasion, the accessibility of solid, brief idleness associations on such an expansive scale may increment the income [11]. The work in [12] highlights the significance of a real-time, large-scale approach for hardware upkeep applications. An IIoT-based energetic generation coordination engineering is displayed in [12] for real-time synchronization of inside and open generation coordination assets. In [13], the optimization of generation planning is based on IIoT-decentralized vitality forecast calculations encouraged by the current state of the machines. Finally, the dynamic decrease of idleness and jitter of the Internet-based network will increment the extent of conceivable applications, as detailed in [14].

The primary module is called Customized Access Module (CA-Module) which is dependable for interfacing heterogeneous gadgets called PMRs (Physical Fabricating Assets) through a gathering of communication conventions. The moment module is called A-Hub (Get to Center) which works as a bridge between manufacturing plant specialist, savvy terminals, and CA-Module through Wi-Fi or Ethernet interfacing and Obliged Application Convention (CoAP) convention. The third and most critical module is called LPS (Neighborhood Pool Benefit) or shrewd terminals. LPSs perform their distinctive capacities and are dependable for information collection, handling, shrewd choice-making, and putting away. Based on information created by PMRs, LPS performs real-time information handling and foresee anticipated generation rate, adds up to vitality utilization and PMRs anticipated upkeep. Each IIHub module is implanted with uncommon reason libraries.

CA-Module has a group of communication protocols which interact with each other using a library called CPPLib (Communication Protocol Package Library). A-Hub has an embedded library called MDIMLib (Multi-dimensional information models library) which help in connectivity. LPS has an embedded library called

Introduction to Industrial Internet of Things (IIoT)

DPALib (Data Processing Algorithm Library) which perform data processing, analysis, and decision making. B. Martinez et al. [15] have proposed an open Industrial hardware platform for sensing and connectivity called I3Mote. The main components of I3Mote include different type of sensors, processor (MSP432), wireless radio interface (CC2650) and multi-source power support (battery, solar and thermoelectric etc.). I3Mote is basically a prototyping hardware which aims to provide all sensing and connectivity features required for IIoT and leads to final product.

CA-Module features a bunch of communication conventions that connected employing a library called CPPLib (Communication Convention Bundle Library). A-Hub has an implanted library called MDIMLib (Multi-dimensional data models library) which offers assistance in the network. LPS has an implanted library called DPALib (Information Preparing Calculation Library) which performs information preparing, investigation, and choice making. Martinez et al. [20] have proposed an open Mechanical equipment stage for detecting and network called I3Mote. Most components of I3Mote incorporate distinctive sort of sensors, processor (MSP432), remote radio interface (CC2650) and multi-source power bolster (battery, sun based and thermoelectric, etc.). I3Mote is essentially prototyping equipment that points to supply all detecting and network highlights required for IIoT and leads to the last item. The first layer is based on SDN whereby the SDN controller is mindful of the administration to arrange synchronizations, control signals, and assets. The moment layer is based on a UAV that serves as a cloudlet. These UAVs give two essential administrations in calamity or crisis circumstances, i.e., information preparation and information communication. At long last, the third layer is based on an RAN which is dependable for radio to get to administrations to the conclusion users. The proposed designs might be utilized in the IIoT environment to supply productive, low-cost information handling, and information communication administrations for open security in emergency/disaster circumstances. Nonetheless, control utilization, issues related to arranging administration, arrangement, direction, and elevation expectation of UAVs are challenges confronted by these proposed architectures [20].

1.4 CHALLENGES OF IIoT

A key reason for receiving IIoT by producers, utility companies, farming makers, and healthcare suppliers is to extend efficiency and productivity through shrewd and inaccessible administration. As an illustration, Thames Water [21], the biggest supplier of drinking and waste-water administrations within the UK, is utilizing sensors, and real-time information securing and analytics to expect hardware disappointments and give a quick reaction to basic circumstances, such as spills or antagonistic climate occasions. The utility firm has as of now introduced more than 100,000 keen meters in London, and it points to cover all clients with keen meters by 2030. With more than 4,200 spills identified on client channels so distant, this program has as of now spared an assessed 930,000 liters of water per day over London. In another case, the arrangement of 800 HART gadgets for real-time prepare administration at Mitsubishi chemical plant in Kashima, Japan has been expanding the generation execution by sparing US$20–30,000 per day that moreover deflected a $3 million shutdown [22]. Accuracy horticulture fueled by IIoT can offer assistance ranchers superior degree rural factors such as soil supplements, fertilizer utilized, seeds planted, soil water, and temperature of putting away create, permitting to screen down to the square foot through a thick sensor sending, subsequently nearly multiplying the efficiency [23–25]. Companies like Microsoft (FarmBeats extend [26, 27]), Climate Corp [28], AT&T [29], and Monsanto [30] are advancing rural IoT. IIoT can moreover altogether affect the healthcare field. In healing centers, human or mechanical mistakes caused by wrong cautions, moderate reactions, and wrong data are still a major reason for preventable passing and quite enduring.

By interfacing distributed medical gadgets utilizing IIoT innovations, clinics can altogether overcome such confinements, in this manner moving forward quiet security and encounters, and more productively utilizing the assets. IIoT moreover gives openings to upgrade effectiveness, security, and working conditions for specialists. For case, utilizing unmanned ethereal vehicles (UAVs) permits reviewing oil pipelines, observing nourishment security utilizing sensors, and minimizing workers' presentation to clamor, and dangerous gasses or chemicals in mechanical situations. Schlumberger, for a case, is presently observing subsea conditions utilizing unmanned marine vehicles, which can travel overseas collecting information for up to a year without fuel or group, moving beneath control created from wave vitality [31]. Through further checking and detecting fueled by IIoT, mining businesses can drastically diminish safety-related occurrences, whereas making mining in cruel areas more prudent and profitable. For the case, Rio Tinto, a driving mining company, serious its computerized operations in Australia to see a more proficient future for all of its mines to diminish the requirement for human mineworkers [32]. Despite the incredible guarantee, there are numerous challenges in realizing the openings advertised by IIoT, which ought to be tended to within the future inquire about. The key challenges stem from the necessities in energy-efficient operation, real-time execution in energetic situations, the requirement for coexistence and interoperability, and keeping up the security of the applications and users' protection.

(a) Security and Privacy

Security is considered a basic concern in IIoT. In common, IIoT could be a resource-constrained communication arrangement which generally depends on low-bandwidth channels for communication among lightweight gadgets concerning CPU, memory, and vitality utilization [33]. For this reason, conventional assurance instruments are not adequate to secure the complex IIoT frameworks, such as secure conventions, lightweight cryptography, and protection affirmation. To secure the IIoT framework, existing encryption methods from mechanical WSNs may be surveyed sometime recently connected to construct IIoT secure conventions. For occasion, rare computing and memory assets anticipate the use of resource-demanding crypto-primitives, e.g., Public-Key Cryptography (PKC).

This challenge is more basic within the applications of enormous information traded with real-time prerequisites. To address protection and security dangers in IIoT, one can contend for an all-encompassing approach as pointed out in [34]. This implies that perspectives such as stage security, secure designing, security administration, character administration, and mechanical rights management must be taken

Introduction to Industrial Internet of Things (IIoT)

under consideration, all through the full life cycle of the frameworks and items. There exist a few security properties to consider when planning a secure IIoT framework: (1) IIoT gadgets ought to be alter safe against potential physical assaults, such as unauthorized re-programming and detached mystery taking whereas permitting the authorized clients to upgrade the security firmware on the gadget. (2) The capacity of IIoT gadget ought to be secured against foe by keeping the information scrambled to keep the privacy. (3) The communication arrange among the IIoT gadgets ought to be secured to keep privacy and judgment. (4) The IIoT foundation needs effective recognizable proof and authorization instruments so that as it was authorized substances can get to the IIoT asset. (5- The framework ought to be accessible inside ordinary operation, indeed with the physical harm to the gadgets by noxious clients. This ensures the strength of IIoT. Ordinarily, symmetric-key cryptography can give a lightweight arrangement for IIoT gadgets. In any case, both the key capacity and the key administration are huge issues in case utilizing symmetric-key encryption, particularly when considering low-capacity gadgets. Also, if one gadget in IIoT is compromised, it may spill all other keys. Public-key cryptography, by and large, gives more secure highlights, and moo capacity prerequisites, but endures from tall computational overhead due to complex encryption. In this way, diminishing the overhead of complex security conventions for public-key cryptosystems remains a major challenge for IIoT security. In PKC, Elliptic-Curve Cryptography (ECC) gives a lightweight arrangement concerning computational assets. It gives a littler key measure, lessening capacity, and transmission necessities. In IIoT frameworks, it is imperative to supply the distinguishing proof to urge the lawful get to. The secure IIoT framework must guarantee the object recognizable proof concerning the keenness of records utilized within the naming frameworks, such as the Space Title Framework (DNS) [35]. The DNS framework can give title interpretation administrations to the web client, in any case, it is in an uncertain way that remains helpless to different assaults by thought enemy. This challenge remains substantial indeed for a bounded and closed environment. Hence, without the keen assurance of the recognizable proof, the complete naming framework is still uncertain. Security expansions to DNS, like, Space Title Benefit Security Expansion (DNSSEC) increments security and is archived in IETF RFC4033 [36]. However, due to its tall computation and communication overhead, it is challenging to straightforwardly apply DNSSEC to the IIoT foundation. IIoT gadgets ought to take after particular plans and rules for verification to exchange/publish their information. Due to the asset limitations of IIoT gadgets, low-cost confirmation plans have not been given as much as required. Even though public-key cryptography frameworks give the strategies for developing confirmation and authorization plans, it comes up short to supply a worldwide root certification specialist (worldwide root CA), which generally ruins numerous hypothetically doable plans from really being sent. Without giving the worldwide root CA, it gets to be exceptionally challenging to plan a secure confirmation framework in IIoT [37]. In this way, right now, on the off chance that we proposed to supply the secure verification for IIoT gadgets, we need to utilize the high-cost arrangements which could be a struggle with the most goal of the lightweight rule of IIoT. In addition, it could be a huge challenge to issue a certification to each protest in IIoT since the overall number of objects can be gigantic. Protection could be an exceptionally wide and different concept. Numerous definitions and viewpoints have been given within the writing. By and large talking, protection in IIoT is the triple ensure for (1) mindfulness of protection dangers forced by things and administrations; (2) person control over the collection and preparing of data; (3) mindfulness and control of consequent utilize and dispersal to any exterior substance [38]. The major challenges for security lie in two viewpoints: information collection preparation and information anonymization handle. Ordinarily, information collection prepares bargains with the collectible information and the get to control to this information amid the information collected from savvy things; information anonymization may be prepared to guarantee information namelessness through both cryptographic security and concealment of information relations. Due to the limitations on the collection and capacity of private data, protection conservation can be guaranteed amid the information collection. Nonetheless, given the differences of the things in information anonymization, different cryptographic schemes may be received which could be a protection challenge. In the meantime, the collected data has to be shared among the IIoT gadgets, and the computation on scrambled information is another challenge for information anonymization [39].

(b) Real-Time Performance

IIoT gadgets are ordinarily sent in loud situations for supporting mission- and safety-critical applications, and have exacting timing and unwavering quality prerequisites on a convenient collection of natural information and appropriate conveyance of control choices. The QoS advertised by IIoT is hence regularly measured by how well it fulfills the end-to-end (e2e) due dates of the real-time detecting and control errands executed within the framework. Time-slotted bundle planning in IIoT plays a critical role in accomplishing the required QoS [40]. For the case, numerous industrial remote systems perform organize resource management using inactive information connect layer planning to attain deterministic e2e real-time communication. Such approaches regularly take an intermittent approach to gather the organized well-being status, and after that recompute and convey the overhauled arrange plan data. This handle in any case is moderate, not adaptable, and brings about impressive organize overhead [41]. The unstable development of IIoT applications particularly in terms of their scale and complexity has drastically expanded the level of trouble in guaranteeing the specified real-time execution. The reality that most IIoT must bargain with startling unsettling influences advance irritates the issue. Startling unsettling influences can be classified into outside unsettling influences from the environment being checked and controlled (e.g., location of a crisis, sudden weight or temperature changes) and inner unsettling influences inside the organized foundation (e.g., connect disappointment due to multi-user impedances or climate-related changes in channel SNR) [42]. In reaction to different inside unsettling influences, numerous centralized planning approaches have been proposed. There are too several works on adjusting to outside unsettling influences in basic control frameworks. For case, rate-adaptive and musical errand models are presented in and, individually, which permit errands to alter periods and relative due dates in a few control frameworks such as car frameworks. Given the prerequisite of assembly e2e due dates, the previously

Introduction to Industrial Internet of Things (IIoT)

mentioned approaches for taking care of unforeseen unsettling influences are nearly all built on a centralized design. Subsequently, most of them have restricted adaptability. The concept of dispersed asset administration isn't unused.

In truth, distributed approaches have been examined reasonably well within the remote arrange community [43]. In any case, these things about regularly are not concerned with real-time e2e limitations. Some, which consider real-time imperatives, primarily center on delicate real-time requirements and don't consider outside unsettling influences that IIoT must get to bargain with. As it were as of late, we have begun to see a few crossover and completely conveyed asset administration approaches for IIoT. Nonetheless, how to guarantee bounded reaction time to handle concurrent unsettling influences is still an open issue [44].

(c) Coexistence and Interoperability

With the quick development of the IIoT network, there will be numerous coexisting gadgets sent in near nearness within the constrained range. This brings forward the imminent challenge of coexistence within the swarmed ISM groups. Hence, impedances between gadgets must be taken care of to keep them operational. Existing and close future IIoT gadgets will most likely have restricted memory and insights to combat impedances or keep it to the least. Whereas there exists much work on remote coexistence considering Wi-Fi, IEEE 802.15.4 systems, and Bluetooth (see studies, they will not work well for IIoT.

Due to their thick and large-scale organizations, these gadgets can be subject to an uncommon number of interferers [45]. Technology-specific highlights of each IIoT innovation may present extra challenges. To guarantee great coexistence it'll end up vital that future IIoT gadgets can identify, classify, and relieve external obstructions. As of late, a few works concerning classifying impedances through range detecting on IIoT gadgets has been displayed but most of the existing work fails since an awfully long examining window is required and the proposed range detecting strategies require much more memory than what is accessible in existing commercial IIoT gadgets. Thus, in a promising strategy was displayed and executed in Crossbow's TelosB mote CA2400 which is prepared with Texas Instrument CC2420 handset [46]. That strategy oversees to classify of outside impedances by utilizing bolster vector machines with a detecting term underneath 300 ms. In addition, existing gadgets based on IEEE 802.15.4 benchmarks don't have any forward mistake redressing (FEC) capabilities to move forward the unwavering quality of the information parcel. There exists a few works that explored mistake control codes for mechanical WSNs and the comes about clearly appear that FEC will move forward with unwavering quality and the coexistence. In any case, most of the accessible FEC strategies are optimized for long bundles. Given that IIoT communication will basically comprise brief parcels (50–70 bytes) and numerous applications are time-critical, more investigation is required to discover great blunder adjusting codes for IIoT communication [47].

If the investigation of blunder rectifying codes for IIoT gadgets ought to be effective, it is additionally imperative that more accentuation be given on exploring and understanding the complex radio environment where numerous of these IIoT gadgets will be conveyed [48]. The quick development of IIoT advances too brings forward the requirements of interoperability. Specifically, within the future, a completely utilitarian advanced biological system will require consistent information sharing between machines and other physical systems from different producers. The need for interoperability among IIoT gadgets will altogether increment the complexity and fetched of IIoT arrangement and integration. The drive towards consistent interoperability will be advance complicated by the long life span of ordinary mechanical hardware, which would require exorbitant retrofitting or substitution to work with the most recent innovations [49]. The challenges of gadget differing qualities in IIoT can be tended to along three measurements: multimode radios, program adaptability, cross-technology-communication. Multimode radios permit different IIoT gadgets to the conversation with each other. Computer program adaptability empowers bolster for numerous conventions, network systems, and cloud administrations [50]. As of late, cross-technology communication without the help of extra equipment has been considered for communication over Wi-Fi, ZigBee, and Bluetooth gadgets. Such approaches are particular to innovations, and hence future investigation is required to empower cross-technology-communication in IIoT gadgets.

(d) Energy Efficiency

Many IIoT applications have to be run for a long time on batteries. This calls for the plan of low-power sensors that don't require battery substitution over their lifetimes. This requests for energy-efficient plans. To complement such plans, upper-layer approaches can play critical parts through the energy-efficient operation. Numerous vitality productive plans for remote sensor arrange (WSN) have been proposed in later a long time, but those approaches are not instantly pertinent to IIoT. IIoT applications regularly require a thick arrangement of various gadgets. Detected information can be sent in a questioned frame or in a nonstop shape which is a thick sending that can devour a critical sum of vitality [51]. Green organizing is thus vital in IIoT to diminish control consumption and operational costs. It'll reduce contamination and outflows and make the foremost of observation and natural preservation. LPWAN IoT innovations accomplish low-power operation utilizing a few energy-efficient plan approaches. To begin with, the more often than not shape a star topology, which kills the vitality expended through bundle steering in multi-hop systems. Moment, they keep the hub plan straightforward by offloading the complexities to the door. Third, they utilize narrowband channels, subsequently diminishing the commotion level and amplifying the transmission run. Even though there are various strategies to realize vitality proficiency, such as utilizing lightweight communication conventions or embracing low-power radio handsets as portrayed over, the later innovation slant in vitality gathering gives another essential strategy to prolong battery life. Thus, vitality gathering could be a promising approach for developing IIoT [52, 53]. For all intents and purposes, energy can be collected from natural sources, be specific, warm, sun-powered, vibration, and remote radiofrequency (RF) vitality sources. Collecting from such natural sources is subordinate to the nearness of the comparing vitality source. In any case, RF vitality gathering may give benefits in terms of

Introduction to Industrial Internet of Things (IIoT)

being remote, promptly accessible within the frame of transmitted vitality (TV/radio broadcasters, portable base stations, and hand-held radios), moo took a toll, and in terms of little shape calculate of gadgets.

1.5 CONCLUSION

The IIoT system allows the industry to collect and analyze a great amount of data that can be used, monetized, and improve the overall performance of the systems for providing new types of services. In this chapter, we have highlighted an overview of the IIoT. Moreover, we took about IIoT architecture and framework. Also, we discussed the challenges that faced the adaption of the IIoT systems in the real world.

REFERENCES

[1] W. Z. Khana, M. H. Rehmanb, H. M. Zangotic, M. K. Afzald, N. Armia, and K. Salahe, "Industrial Internet of Things: recent advances," *Enabling Technologies and Open Challenges*, Elsevier, 2019, https://www.researchgate.net/publication/337498167.

[2] L. Da Xu, Wu He, Shancang Li, "Internet of things in industries: a survey 2013," http://www.ieee.org/publications_standards/publications/rights/index.html for more information.

[3] T. Qiu, J. Chi, X. Zhou, Z. Ning, M. Atiquzzaman, and D. Oliver Wu, *Edge computing in industrial internet of things: architecture, advances and challenges*, Newcastle University, 2020 at 11:09:04 UTC from IEEE Xplore.

[4] E. Sisinni, A. Saifullah, S. Han, U. Jennehag, and M. Gidlund, "Industrial Internet of Things: challenges, opportunities, and directions," *IEEE Transactions on Industrial Informatics*, vol. X, no. X, 2018, IEEE. doi:10.1109/TII.2018.2852491.

[5] M. R. Palattella, M. Dohler, A. Grieco, G. Rizzo, J. Torsner, T. Engel, and L. Ladid, "Internet of Things in the 5G era: enablers, architecture, and business models," *IEEE Journal on Selected Areas in Communications*, vol. 34, no. 3, pp. 510–527, 2016.

[6] D. Bandyopadhyay and J. Sen, "Internet of Things: applications and challenges in technology and standardization," *Wireless Personal Communications*, vol. 58, no. 1, pp. 49–69, 2011.

[7] M. R. Palattella, P. Thubert, X. Vilajosana, T. Watteyne, Q. Wang, and T. Engel, *Internet of Things. IoT Infrastructures: Second International Summit*, 2016.

[8] L. D. Xu, W. He, and S. Li, "Internet of Things in industries: a survey," *IEEE Transactions on Industrial Informatics*, vol. 10, no. 4, pp. 2233–2243, 2014.

[9] M. Wollschlaeger, T. Sauter, and J. Jasperneite, "The future of industrial communication: automation networks in the era of the Internet of Things and industry 4.0," *IEEE Industrial Electronics Magazine*, vol. 11, no. 1, pp. 17–27, 2017.

[10] W. He and L. Xu, "A state-of-the-art survey of cloud manufacturing," *International Journal of Computer Integrated Manufacturing*, vol. 28, no. 3, pp. 239–250, 2015.

[11] I. Lee, "An exploratory study of the impact of the Internet of Things (IoT) on business model innovation: building smart enterprises at fortune 500 companies," *International Journal of Information Systems and Social Change (IJISSC)*, vol. 7, no. 3, pp. 1–15, 2016.

[12] T. Qu, S. P. Lei, Z. Z. Wang, D. X. Nie, X. Chen, and G. Q. Huang, "Iot-based real-time production logistics synchronization system under smart cloud manufacturing," *The International Journal of Advanced Manufacturing Technology*, vol. 84, no. 1, pp. 147–164, 2016.

[13] S. G. Pease, R. Trueman, C. Davies, J. Grosberg, K. H. Yau, N. Kaur, P. Conway, and A. West, "An intelligent real-time cyber-physical, toolset for energy and process prediction and optimisation in the future industrial Internet of Things," *Future Generation Computer Systems*, vol. 79, pp. 815–829, 2018.

[14] T. H. Szymanski, "Supporting consumer services in a deterministic, industrial internet core network," *IEEE Communications Magazine*, vol. 54, no. 6, pp. 110–117, 2016.

[15] S. Lin, B. Miller, J. Durand, G. Bleakley, A. Chigani, R. Martin, B. Murphy, M. Crawford, "The industrial Internet of Things volume g1: reference architecture," *Industrial Internet Consortium*, pp. 10–46, 2017.

[16] G. Campobello, M. Castano, A. Fucile, A. Segreto, *Weva: a complete solution for industrial Internet of Things*, In: *International Conference on Ad-Hoc Networks and Wireless*, Springer, pp. 231–238, 2017.

[17] C. K. M. Lee, S. Z. Zhang, K. K. H. Ng, "Development of an industrial Internet of Things suite for smart factory towards re-industrialization," *Advances in Manufacturing*, vol. 5, no. 4, pp. 335–343, 2017. doi:10.1007/s40436-017-0197-2.

[18] W. Z. Khan, M. Y. Aalsalem, M. K. Khan, M. S. Hossain, M. Atiquzzaman, *A reliable Internet of Things based architecture for oil and gas industry*, In: *Advanced Communication Technology (ICACT), 2017 19th International Conference on*, IEEE, pp. 705–710.

[19] F. Tao, J. Cheng, Q. Qi, "IIHub: an industrial internet-of-things hub toward smart manufacturing based on cyber-physical system," *IEEE Transactions on Industrial Informatics*, vol. 14, no. 5, pp. 2271–2280, 2018, doi:10.1109/tii.2017.2759178

[20] B. Martinez, X. Vilajosana, I. Kim, J. Zhou, P. Tuset-Peir´o, A. Xhafa, D. Poissonnier, X. Lu, "I3mote: an open development platform for the intelligent industrial internet," *Sensors*, vol. 17, no. 5, 2017.

[21] Z. Kaleem, M. Yousaf, A. Qamar, A. Ahmad, T. Q. Duong, W. Choi, A. Jamalipour, "Uav-empowered disaster-resilient edge architecture for delay-sensitive communication," *IEEE Network*, pp. 1–9, 2019. doi:10.1109/MNET.2019.1800431.

[22] J. P. Tomas, *Thames water rolls out smart meter project in london*, 2017, https://wiprodigital.com/cases/progressive-metering-a-utilitys-strategic-move-into-predictive-planning/.

[23] http://en.hartcomm.org/hcp/tech/applications/applicationssuccessmitsubishichemical.html.

[24] M. H. Almarshadi and S. M. Ismail, "Effects of precision irrigation on productivity and water use efficiency of alfalfa under different irrigation methods in arid climates," *Journal of Applied Sciences Research*, vol. 7, no. 3, pp. 299–308, 2011.

[25] H.-J. Kim, K. A. Sudduth, and J. W. Hummel, "Soil macronutrient sensing for precision agriculture," *Journal of Environmental Monitoring*, vol. 11, no. 10, pp. 1810–1824, 2009.

[26] N. D. Mueller, J. S. Gerber, M. Johnston, D. K. Ray, N. Ramankutty, and J. A. Foley, "Closing yield gaps through nutrient and water management," *Nature*, vol. 490, no. 7419, pp. 254–257, 2012.

[27] D. Vasisht, Z. Kapetanovic, J. Won, X. Jin, R. Chandra, S. Sinha, A. Kapoor, M. Sudarshan, and S. Stratman, *Farmbeats: an iot platform for data-driven agriculture*, In: *14th USENIX Symp. on Net. Syst. Design and Implementation (NSDI)*, pp. 515–529, 2017.

[28] Microsoft, "FarmBeats: IoT for agriculture," https://www.microsoft.com/en-us/research/project/farmbeats-iot-agriculture/.

[29] C. Corporation, "Data-driven agricultural decisions and insights to maximize every acre," https://www.climate.com.

Introduction to Industrial Internet of Things (IIoT) 17

[30] AT&T M2X, "Agriculture iot software as a service (saas)," https://m2x.att.com/iot/industry-solutions/iot-data/agriculture/.

[31] J. Hawn, "Agricultural iot promises to reshape farming," *RCR Wireless News*, November 2015, https://www.rcrwireless.com/20151111/internet-of-things/agricultural-internet-of-things-promises-to-reshape-farming-tag15.

[32] Schlumberger, "Schlumberger robotics services," http://www.slb.com/services/additional/robotics-services.aspx.

[33] T. Heer, O. Garcia-Morchon, R. Hummen, S. L. Keoh, S. S. Kumar, and K. Wehrle, "Security challenges in the ip-based Internet of Things," *Wireless Personal Communications*, vol. 61, no. 3, pp. 527–542, 2011.

[34] A. W. Atamli and A. Martin, *Threat-based security analysis for the Internet of Things*, In: *International Workshop on Secure Internet of Things (SIoT)*, IEEE, pp. 35–43, 2014.

[35] Z.-K. Zhang, M. C. Y. Cho, C.-W. Wang, C.-W. Hsu, C.-K. Chen, and S. Shieh, *Iot security: ongoing challenges and research opportunities*, In: *IEEE 7th International Conference on Service-Oriented Computing and Applications (SOCA)*, pp. 230–234, 2014.

[36] R. Arends, R. Austein, M. Larson, D. Massey, and S. Rose, "Dns security introduction and requirements," *Tech. Rep.*, 2005.

[37] G. Baldini, T. Peirce, M. Botterman et al. "Iot governance, privacy and security issues," *Position paper, European Research Cluster on the Internet of Things*, 2015.

[38] S. Raza, Lightweight security solutions for the Internet of Things, Ph.D. dissertation, Malardalen University, V ¨ aster ¨ as, Sweden, 2013.

[39] J. H. Ziegeldorf, O. G. Morchon, and K. Wehrle, "Privacy in the Internet of Things: threats and challenges," *Security and Communication Networks*, vol. 7, no. 12, pp. 2728–2742, 2014.

[40] P. Ferrari, A. Flammini, E. Sisinni, D. Brando, and M. Rocha, "Delay estimation of Industrial IoT applications based on messaging protocols," *IEEE Transactions on Instrumentation and Measurement*, pp. 1–12, 2018.

[41] S. Han, X. Zhu, D. Chen, A. K. Mok, and M. Nixon, *Reliable and real-time communication in industrial wireless mesh networks*, In: *Proceedings of IEEE Real-Time and Embedded Technology and Applications Symposium (RTAS)*, pp. 3–12, 2011.

[42] O. Chipara, C. Lu, and G.-C. Roman, "Real-time query scheduling for wireless sensor networks," *IEEE Transactions on Computers*, vol. 62, no. 9, pp. 1850–1865, 2013.

[43] J. Kim, K. Lakshmanan, and R. Rajkumar, *Rhythmic tasks: a new task model with continually varying periods for cyber-physical systems*, In: *IEEE/ACM Third International Conference on Cyber-Physical Systems (ICCPS)*, pp. 55–64, 2012.

[44] T. Zhang, T. Gong, Z. Yun, S. Han, Q. Deng, and X. S. Hu, *Fd-pas: a fully distributed packet scheduling framework for handling disturbances in real-time wireless networks*, In: *IEEE Real-Time and Embed. Tech. and App. Symp. (RTAS)*, pp. 1–12, 2018.

[45] D. Yang, Y. Xu, and M. Gidlund, *Coexistence of ieee802.15.4 based networks: a survey*, In: *Proceedings of the 36th Annual Conference on IEEE Industrial Electronics Society (IECON)*, pp. 2107–2113, 2010.

[46] T. M. Chiwewe, C. F. Mbuya, and G. P. Hancke, "Using cognitive radio for interference-resistant industrial wireless sensor networks: an overview," *IEEE Transactions on Industrial Informatics*, vol. 11, no. 6, pp. 1466–1481, 2015.

[47] F. Barac, M. Gidlund, and T. Zhang, "Ubiquitous, yet deceptive: hardware-based channel metrics on interfered WSN links," *IEEE Transactions on Vehicular Technology*, vol. 64, no. 5, pp. 1766–1778, 2015.

[48] F. Barac, S. Caiola, M. Gidlund, E. Sisinni, and T. Zhang, "Channel diagnostics for wireless sensor networks in harsh industrial environments," *IEEE Sensors Journal*, vol. 14, no. 11, pp. 3983–3995, 2014.

[49] L. Ascorti, S. Savazzi, G. Soatti, M. Nicoli, E. Sisinni, and S. Galimberti, "A wireless cloud network platform for industrial process automation: critical data publishing and distributed sensing," *IEEE Transactions on Instrumentation and Measurement*, vol. 66, no. 4, pp. 592–603, 2017.

[50] S. M. Kim and T. He, *Freebee: cross-technology communication via free side-channel*, In: *Proceedings of the 21st Annual International Conference on Mobile Computing and Networking*. ACM, pp. 317–330, 2015.

[51] A. Saifullah, M. Rahman, D. Ismail, C. Lu, R. Chandra, and J. Liu, *SNOW: sensor network over white spaces*, In: *The 14th ACM Conf. on Embedded Network Sensor Systems (SenSys)*, pp. 272–285, 2016.

[52] T. Rault, A. Bouabdallah, and Y. Challal, 'Energy efficiency in wireless sensor networks: a top-down survey," *Computer Networks*, vol. 67, pp. 104–122, 2014.

[53] 3GPP, "Standardization of NB-IOT completed," 2016, http://www.3gpp.org/news-events/3gpp-news/1785-nb iot complete.

2 Challenges in Industrial Internet of Things (IIoT)

Sarbagya Ratna Shakya

University of Southern Mississippi, Hattiesburg, USA

Sudan Jha

CHRIST University, Delhi-NCR Campus, Delhi, India

CONTENTS

2.1 Introduction 20
2.2 Application of IIoT 21
    2.2.1 Manufacture Industry 22
        2.2.1.1 Sectors Where IIoT Has Been Adopted in the Manufacturing Industry 22
        2.2.1.2 Challenges for IIoT in Manufacturing 24
    2.2.2 Agriculture Industry 25
        2.2.2.1 Sectors That Implement IIoT in Agriculture Industry 26
        2.2.2.2 Challenges for IIoT in the Agriculture Industry 27
    2.2.3 IIoT in Connected Logistics, Transportation, and Warehousing 28
        2.2.3.1 Sectors That Implement IIoT in Connected Logistics and Transportation 28
        2.2.3.2 Challenges of IIoT in Logistics and Transportation 29
    2.2.4 IIoT in Healthcare 30
        2.2.4.1 Sectors That Implement IIoT in Healthcare 30
        2.2.4.2 Challenges of IIoT in Healthcare 31
2.3 Challenges Based on IIoT Components 31
    2.3.1 Challenges in Devices 31
    2.3.2 Challenges in Network 32
    2.3.3 Challenges in Data 34
2.4 Future Technology and Its Challenges 35
    2.4.1 5G-based IIoT 35
    2.4.2 Blockchain-Based IIoT 36
2.5 Conclusion 37
Bibliography 37

DOI: 10.1201/9781003102267-2

2.1 INTRODUCTION

With the advanced use of sensors, smart devices, and low-cost communication devices, the use of the Internet of things (IoT) has been increasing in recent years [1]. In IoT, the sensors, actuators, and other static and mobile devices are interconnected and communicated through the Internet, reducing human involvement, and increasing efficiency. The extension of IoT, applied in industrial sectors with machine-to-machine communication, big data, automation, robotics, and machine learning leads to the development of the Industrial Internet of Things (IIoT). The objective of IIoT has been to converge the manufacturing physical space and manufacturing cyberspace with the characteristics of automation, smart interconnection, real-time monitoring, and collaborative control.

The integration or convergence of information technology (IT) with operational technology (OT) such as operational networking and industrial control systems has made industries possible to have better efficiency, reliability in their operations, better system integration in automation and optimization, and better visibility in their supply chain and logistics. IT is primarily responsible for handling computing and data processing. It will handle the data from storing to securing. OT is responsible for managing and controlling industrial operations like productivity, people, and machinery. IIoT brings these two previously separated sectors in a single line. The amounts of data generated from IIoT sensors, software, and large assets can be used to improve internal processes, increase productivity, and make business decisions. The use of smart sensors and actuators in industries provides real-time data. These real-time data can be analyzed and provide valuable information to make decision-making easier and take specific actions. This allows industries to detect errors or inefficiencies much earlier and take necessary actions immediately for the maintenance process. Also, using the data provided through the communication between machines can help to predicts faults in the machine before the time they occur. This will help in fault detection and maintenance of the system much earlier leading to the efficiency of the system.

When looking at the history of evolution in the industry, the first industrial revolution started with the development of water, steam, and coal-powered machines (Industry 1.0). This helped the worker, and the production capabilities increased. The second industrial revolution was considered with the invention of electricity. It was used to replace water and steam engines to make more efficient and effective manufacturing facilities. Further, employees overloaded with specific tasks in order to speed up the production process, optimizing the human effort and workplaces were the major agenda of '*mass- production line*' (Industry 2.0). The third industrial revolution started with the development and use of electronic devices such as the transistor, integrated circuits, that helped to develop fully automatic systems, controlled by software systems. Also, the geographic dispersion of manufacturing factories to low-cost countries developed the concept of supply chain management and developed the IT-enabled manufacturing plants (Industry 3.0). After that, the fourth industrial revolution is considered with the advancement of the IoT that implement technologies such as robotics, machine learning, artificial intelligence, advanced materials, and augmented reality and data science to analyze, guide, and share the information that leads to the smart factories (Industry 4.0). Its main research domains include cyber-physical systems, the IoT, cloud computing, and cognitive computing for automation and data exchange in manufacturing technologies [2]. The implementation of IoT in the industry should fulfill some of the characteristics of Industry 4.0 such as decentralization, interoperability, virtualization, real-time response, and modularity. The benefits it should generate can be listed as increased productivity, improved business continuity, better working conditions, customization, and agility [3].

The industrial IoT is expected to transform the industry through intelligent interconnect objects that dramatically improve performance, lowering down the operating costs, with increased reliability and transmit and process massive data collected from the manufacturing process, it requires high-speed, low-latency, and high-reliability communication technology [4, 5].

2.2 APPLICATION OF IIoT

The application of IIoT has open opportunities in several industries. Some of the industrial markets for implementing IIoT are aerospace and defense, automotive, agriculture, energy & utilities, healthcare, manufacturing industry, transportation, and many others. According to the research based on data gathered from the company's customer base, published by PTC [6], one of the leading IIoT software platform providers, the leading sectors deploying IIoT solution were industrial product (25%), followed by electronics and high-tech industry (23%), automotive 13%, and retail and consumer (8%). The details of the survey are shown in Figure 2.1. The main objective of implementing IIoT in different sectors has been mainly to reduce cost, optimize the supply chain, and comply with regulatory requirements, providing agility to disruptive business models, stronger demand for transparency outside of the organization, and many more.

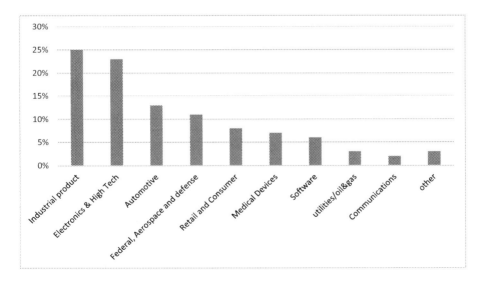

FIGURE 2.1 Industrial IoT adoption by industry [6].

22   Industrial Internet of Things

Here we discuss some of the key sectors that have shown potential and growth with the adoption of IIoT and some of the challenges it faced, categorized based on the industry.

2.2.1   MANUFACTURE INDUSTRY

This is the largest sector in IIoT applications, based on the investment done in recent years. According to research published by meticulous market research Pvt Ltd [7], the industrial IoT market is expected to grow at a CAGR of 16.7% from 2019 to 2027 to reach around $263.4 billion by 2027. With the increase in the demand for the product and to meet consumer expectations, many manufacturing industries are responding to this by implementing smart IIoT technology in their manufacturing operations. The industrial revolution, use of data across the entire manufacturing value chain, the digital transformation, and automation in the manufacturing industry have made it possible to increase efficiency and improve manufacturing operations, product management, and maintenance and service. The use of robotics, more smart automation backed by big data analytics, and the integration of IT, control systems, and OT has proved to be key factors for IIoT to be popular in the manufacturing industry. The ability to predict future events using real-time insights from the system has helped for quick decision-making, fostering streamlined maintenance, asset monitoring, and increase efficiency and production. Many initiatives like Made in China 2025 and Germany's Industry 4.0[8] have been put forwarded to encourage manufacturing enterprises to upgrade factories to become more competitive, innovative, and efficient [9].

2.2.1.1   Sectors Where IIoT Has Been Adopted in the Manufacturing Industry

a.   Real-time Assets Monitoring and Predictive Maintenance

Using IIoT, machines connected with systems in the manufacturing industry can provide real-time data and enables it to monitor the machines in real time. The sensors track process, monitor the assets, and send the data for decision-making for manufacturing and maintenance decisions. This can also be applied for remote manufacturing where assets can be monitored remotely to maintain reliability, security, and safety. A real-world example is US-based independent AMR solutions that deploy smart meter monitoring systems that use IIoT to monitor the smart meters in real time. This has helped them to keep records of water consumptions, leakage, wastage, theft, meter health, and has helped in saving billions of gallons of water, and accurate billing which has improved their revenue collections.

Redictive maintenance estimates the asset failure and other related issues by analyzing production data. For a single targeted machine part of the whole production line, the organization uses IIoT to know beforehand. It helps the organization to avoid unnecessary labor costs for preventative maintenance as IIoT technology will predict and inform when the service is needed. Also, the equipment can sustain damage over time which can be monitored by data from temperature sensors,

Challenges in Industrial Internet of Things (IIoT) 23

vibration sensors, which can cause the problem to the operation condition. Also, data such as temperature, humidity, raw materials used, other atmospheric conditions in the working environment, the composition of the materials used, and the impact of the shipping container/vehicle's environmental condition have had or may have had on the product can be used to assure the quality control of the product. This will help to reduce cost by decreasing machine downtime and increasing operational efficiency.

b. Inventory management and production flow

IIoT uses software to track inventory and monitor events across the supply chain and prevents errors in inventory management. The information about realistic measures, estimation of materials, and alert system or notification about important requirements or changes will help the inventory manager. For example, a shelf level indicator could be used to order the supply when the item falls below a certain threshold. This can reduce human error while inventory check and all this process can be done automatically. This can also be used to monitor the process in the production flow. It can point out lags, errors, or ways to improve the production line by analyzing the data coming from the sensors in the production line in real time. This can reduce cost, prevents slowdowns, and removes discontinuation of production. It can also provide performance indicators of the machines and working conditions and provide safety and health conditions of the employee to reduce injury and accidents in the production line.

c. Reduce operational cost

With the increase in the competition in the market, the manufacturing company is looking for ways to reduce their operational cost as there may be a thin margin in their profits. Reducing costs in the production process can be a good way to increase the gap in their profits. As even a small reduction in the consumption of any materials in the industry can reduce a greater cost of their expenses and hence saved a higher amount which leads to higher profit margins. For example, a small reduction in the consumption of electricity in the manufacturing industry can save a million dollars per year which can be obtained by improving the efficiency of the motor-driven systems of the factories. Hence, smart metering can be used to track the consumption of resources such as fuel, water, and electricity. This information can help operational managers to know exactly how much resources have been consumed, where are they being consumed, and through effective planning and management, how to decrease consumption and reduce operational cost. Energy thefts track renewable power and remotely managing manual operations are some of the other factors to reduce operational cost.

d. Supply chain visibility

The industrial IoT is playing a growing role in wider supply-chain operations. The users of sensors within a manufacturing supply chain provide insights into the supply chain creating operational visibility to improve performance. The manufacturing digitization and cloud-enabled technologies have been driving end-to-end supply chain to visibility. Its application helps to reduce miscommunications, reduce time waiting for assets and between operations, and detect its cause, impact, and ways to reduce it. The sensors placed on assets can help software to pinpoint the location of the asset and identify the location anywhere which extends operational visibility to all parts of the manufacturing organizations, suppliers, and end-users. This can help reduce error and made it more efficient. It also supports regulatory compliance, giving all the information about the component, where it has come from, who handled them, and the condition they were in. The information about product production, transported, stored conditions, will help to prevent mislabeling, and help combat counterfeit medications. It can provide beneficial applications in complex and multinational supply chains to remotely locate the asset and view and track the logistics flow throughout the supply chain.

2.2.1.2 Challenges for IIoT in Manufacturing

a. Security

One of the features of the implementation of IIoT is the availability of data from the sensors, actuator, and manufacturing plants that can be analyzed to improve effectiveness and efficiency and increase production. Security concerns about these data, which are mostly stored in clouds, in terms of cyber-attack or data breaches, how they are transmitted and analyzed arise along with the risk of the loss of intellectual property and trade secrets. With more smart technology being implemented, and the production is more technology-driven, the challenges created by cybersecurity threats escalate. Also, if the network can be accessed, hackers can gain access to the machines and disrupt the production and damage it for a long period. So, any machine, device, and asset that has been connected to the Internet or network and is being controlled through it can be hacked and attacked. It becomes a challenge to maintain trust between the consumer and the service provider regarding data management and data security although encryption of transmitted data, putting a standard firewall, and upgraded network securing protocols are some methods used for data protection and network security.

b. Cost of Integration.

With a long history of the manufacturing industry, the use of traditional machines and their legacy systems have been there for many years. Most of the existing machines are running perfectly and smoothly and the system is also used to and comfortable with these techniques. This old equipment would not be compatible with the IIoT interface and may not integrate easily with this technology. The high cost of deploying the new machines and processes may make it hard to convince the manufacturer to replace these old machines without expensive capital purchases. These machines should be equipped with sensors for two-way interaction and useful information should be extracted from the data collected from these sensors. The cost for

Challenges in Industrial Internet of Things (IIoT) 25

analyzing these data, storage resources, and computing power will also increase the expenditure of the industry.

c. Lack of Standards

Lack of standards and protocol is another challenge for a highly growing industry to meet coherent security progress. There is a need to develop a standard and protocols for all the development to self-regulate and make it more compatible with the local protocols. The connection between devices and machines that are using different protocols and architecture and to transmit it more securely with cross-platform connections with IT standards has been a great challenge for the IIoT industry.

d. Connectivity

The reliability of the product greatly depends on the machine to machine (M2M) and machine to human connections. Mostly in industry, wired connections were preferred but in recent times wireless connection is more preferred in IIoT technology which creates a redesign of a new network infrastructure. High-speed Internet and reliable connection network have been a key factor for the success of the implementation of IIoT in the manufacturing industry. But with wireless networking, the challenges it brings toward the security of the connection network cannot be ignored.

e. Skills

With this relatively new technology but advancing so rapidly, the need for highly skilled manpower to design, develop, implement, and maintain an IIoT in the production line to increase manufacturing efficiency in the industry increases. From system architect to hardware engineer, along with data analysts, this highly skilled manpower is needed to better understand, implement, and select the hardware to use within the system and gain benefits from implementing IIoT technology in the industry. The challenges will be to provide this skilled manpower for the vast manufacturing industry adopting IIoT globally in near future.

2.2.2 Agriculture Industry

With the global human population increasing more than 1.1% every year, to fulfill the demand for food, the agriculture industry will be an important field in the future. Many factors such as climate change, global warming, scarcity of water, limited fertile land, and other variables will make it more challenging to meet the demand for food for the growing population. To find the solution for this, and to increase the production to meet this demand, the agriculture industry is looking at IIoT where they can use data-driven technology and smart devices to improve their agriculture practices.

IIoT has been looked like a way to improve traditional farming practices. The real-time data collected from the smart sensors can provide useful information about different factors: such as the prediction of rainfall and soil fertility conditions which can be beneficial for increasing productivity. The use of technology can also make farming easier. Companies such as Microsoft [10, 11], climate crop [12], AT&T, and Monsanto [13] are promoting agriculture IoT. The implementation of automation along with Artificial Intelligence has helped the farmers use the farming equipment more efficiently. For example, simple use of a water tank monitoring system in real-time can implement the irrigation process more efficiently. Also adopting these technologies in agriculture has helped farmers to take agricultural decisions, backed by smart data analysis to increase the production efficiently and more effectively with a decrease in the cost. Efficient use of resources, manpower, and information about the real-time situation has been a major advantage of implementing these technologies, but the lack of a constant and reliable communication network has been a major challenge to implement IIoT in remote places.

2.2.2.1 Sectors That Implement IIoT in Agriculture Industry

a. To monitor climate conditions

Climate conditions are the most important factors in increasing agricultural productivity. Climate can very much deteriorate the quantity and quality of production. With the help of IIoT and sensors that are placed in the field, real-time weather conditions such as temperature, humidity can be monitored. These real-time data provided have been used to predict the rainfall and weather conditions more precisely and accurately in the future for making agricultural-based decisions and taking precautions to reduce damage and reap more agricultural benefits.

b. Livestock management

Not only in a crop, but IIoT can also be and has been implemented in livestock management. The data can be used to improve the production of meat, dairy, and genetic production. Also, to monitor the health of livestock, disease management, vaccines, and genetics production, the use of this technology has been beneficial. Monitoring environmental changes to reduce risks of the environment in the livestock is any sector that has helped farmers. For example, to monitor the health of aquatic animals due to environmental factors, automation of feeding systems can help reduce production costs and increase production.

c. Smart irrigation

Water is the most important and essential commodity in agriculture. Traditional irrigation methods involve surface irrigation, flood irrigation, or manual spraying of water. These methods can cause overwatering which can be a great disadvantage within limited water supply conditions. Also, this can ruin crops and can have adverse risks in the production. The implementation of IIoT in irrigation has helped to smartly manage water consumption by using modern irrigation methods such as micro-sprinklers, drip lines, and central pivots. These smart irrigation methods can be

Challenges in Industrial Internet of Things (IIoT)

monitored and controlled remotely and can also be automated according to the requirement of the soil type.

d. Precision farming

Precision farming uses automated devices and IIoT devices and smart sensors for data collection, data analyzing, and use the gathered information for quick and efficient decision-making to increase efficiency, productivity, and sustainability. Livestock monitoring and management, irrigation management, vehicle tracking, field observation, and inventory monitoring are some of the sectors where applications of precision farming have been proven beneficial to increase productivity and to make farming more precise, automatic, and controlled. Examples include smart crop sensors to monitor the hydration, pH value, and nutrient levels of the soil, drones for aerial surveillance, fertilizing and spraying, robotics for autonomous milking, and autonomous tractors to work in tandem to till seed, and plant, connected devices to detect and control water flow, and nutrient level and lessens water waste.

e. Greenhouse

With a smart greenhouse, IIoT has been able to automatically adjust the required temperature inside the greenhouse. This has made the entire process automatically without any human intervention. This has reduced cost and increase accuracy by maintaining a certain climate condition inside the greenhouse. The sensors connected in the greenhouse will give information such as pressure, humidity, temperature, and light levels which will then be used to control these systems more precisely and in real-time.

2.2.2.2 Challenges for IIoT in the Agriculture Industry

a. Communication Infrastructure

For IIoT applications to be more effective in farming, farmers need a more efficient network with high bandwidth, low-latency network to support high stream data and videos communicated through sensors and monitoring devices. A M2M connectivity with constant communication and connectivity guaranteed within the coverage area between these devices, despite the mobility of the sensor and the devices, should adapt to change in network topology and environmental conditions. Hence, the challenge will be to provide reliable connectivity in the field and remote areas of farming for the adoption of IIoT in the farming industry.

b. Farmer Modernization

Study shows that most of the farmers are aging now. The digital skills of the workforce are limited and thus will decrease their ability to invest and modernize their practices of production. Further investment in the new tools, new technology, and training to adopt these technologies has been an essential challenge in the adoption of this technology in the farming industry.

c. Lack of Standards

With multiple manufactures, devices, and equipment, the development of technical standards has been a great challenge for the implementation of IIoT in agriculture. The standards should ensure that the new devices are compatible over time and can link the different systems and devices together in a unified system covering all the aspects of the agricultural exploitation [14].

2.2.3 IIoT in Connected Logistics, Transportation, and Warehousing

The transportation, logistics, and warehouse sectors are the second largest market for IIoT based on financial investment. The growing emergence of digital supply chain and connected logistics reality is to increase the value chain with advanced communication and monitoring systems and to achieve 7Rs of logistics [15], getting the right product to the right customer at the right time, right place, and right condition in the right quantity and at the right cost. The data collected from IIoT can be implemented and applied for better security, faster shipping, enhanced supply chain management, and improved process. With IIoT, the connections between M2M will enable communication between vehicles, packages, containers, loading equipment, and other devices throughout the supply chains. This can minimize cargo theft, economic loss, and human loss. The use of sensors, telematics, and smart devices can help track the vehicles throughout the supply chain from the time it leaves the factory to the time it reaches its destination. This can also track the lost or stolen property and will have a greater chance of recovery. Thus, the connected logic market consists of services, security, the cloud, and big data analysis.

2.2.3.1 Sectors That Implement IIoT in Connected Logistics and Transportation

a. Vehicle and Equipment maintenance

Fleet management and maintenance can be automated using IIoT to check cargo vehicles' condition like stability control, tire pressure, and coolant level. The breakdowns and unexpected failures can be prevented by using an app-controlled maintenance alert system and having real-time access to these data make the maintenance of these problems easy and effective. This will also ensure the safety of the drivers and transportation personnel working on this vehicle. Sensors like GPS and RFID [16] help to track and monitor the vehicles in real time. Not only the vehicles, but the smart devices can also learn the driver's behaviors, vehicle performance, vehicles condition like accelerating and decelerating, and can send real-time alerts to the fleet manager and back to the drivers in real time. The information like the speed of the vehicles, driving habits of the driver, and the time between the stops can be useful to teach drivers how to drive more efficiently and reduce fuel consumption. The data

Challenges in Industrial Internet of Things (IIoT)   **29**

coming out from the sensors connected to the fleet can also be analyzed to calculate the optimized route, fuel efficiency, and can manage route management automatically. The coordinate logistic events like pick-up and drop-off along the supply chain can be managed which will benefit from the warehouse employee to the transportation employee to the end-users. For delivery of the food and medicines, the sensors and devices can be used for maintaining the right temperature and humidity condition inside the cargo truck. The real-time information coming from these sensors can be used to monitor and correct it if there are any problems. The delivery speed can also be increased along with good pricing with a high-level of coordination and precision.

b. Inventory tracking and warehouse management

IIoT can be used in the industry to manage and track inventory and equipment. The company has been able to track inventory items, monitor their status, and create a smart warehouse system. Information such as if an excessive amount of goods is stored in the storage or there is a low supply of important items can be determined in real time with the sensors used in the inventory without any human involvement. This can be achieved and can track inventory items, monitor their status, and create a smart warehouse. This helps an employee to efficiently locate the product, minimize human errors, and reduce losses. Besides inventory tracking and management, effective management of the warehouse can be achieved to increase the productivity of the organization and help in the smooth transportation of the goods swiftly and at an inappropriate time. This can reduce manual conducted routine operations, and the data that it provides can help deliver the product in time.

2.2.3.2 Challenges of IIoT in Logistics and Transportation

a. The internal struggle between operational management and IT teams

Purchasing IIoT devices for the connected supply chain can disrupt the operational management and IT teams. Technology keeps changing and the need to upgrade or update the latest technology can increase capital investment which operational management may not agree with. Most of the Operational management accustomed to purchasing technology that is durable for at least a decade which is not possible. The adoption of new IIoT should make a bigger impact on the process or solve the problem existing in the supply chain. For example, replacing barcode scanners with RFID technology and cloud-based GPS sensors can increase the efficiency in tracking but they should be convinced about the flexibility of these devices about what happens when a better RFID sensor or better GPS tracking software available in a year.

b. Data Analysis

One of the biggest advantages of adopting IIoT is the availability of a large amount of data from the sensors used. Raw data collected from the sensors throughout the production line, supply chain, machines, vehicles, and products may be incomplete, insufficient, or inaccurate. Digital data coming from other sources like offline

30 Industrial Internet of Things

processes, partners, social media, traditional ERP systems all need to properly account for and managed to gain useful insights from the data. This requires a quality analytics process, the right people, technology, and resources to capitalize on the data's full value. Data management functions, data integration, data quality, and data governance should be used to handle all the data collected from the supply chain to inspire that make a fast, well-informed decision and fix potential issues.

2.2.4 IIoT in Healthcare

Another important sector where the application of IIoT has been increasing in recent years is the healthcare sector. The increase in demand of the rising aging population with chronic disease has made the healthcare system difficult to manage [17]. Gartner and Forbes have estimated that up to 2020, the IoT will contribute $1.9 trillion to the global economy and $117 billion to the IoT-based healthcare industry [18]. As more and more medical devices, applications, and technologies are connected through the Internet and Bluetooth, this communication along with the addition of cloud computing [19] can greatly transform the healthcare sectors. The doctors and healthcare providers will have immediate access to the health information of the patient in real-time which can help them to make decisions regarding the patients' health condition and treatment. The industrial IoT sensors for healthcare can be used in applications such as healthcare monitoring systems, a wearable which when connected to the Internet or cloud can provide information about the health condition of the patients in real time. This will enable continuous monitoring of patients remotely. Also, telehealth, automatic healthcare system, and mobile health alert systems can be implemented with the adoption of IIoT in healthcare. This will also help to manage the asset, better clinical efficiency during surgeries and emergencies, and will enhance the health and safety of the patient.

2.2.4.1 Sectors That Implement IIoT in Healthcare

a. Remote healthcare

This enables healthcare workers to monitor, treat, and give guidelines to a patient from any location. The portable devices, home automated machines, apps with embedded connectivity are capable to monitor the various parameters related with the health such as blood pressure, glucose level, and provides health indicators to health officials from which they will be able to monitor and react to the data in real-time. Smart pill dispensers can be used to monitor patient's recommended dosage instructions. With this doctor may be able to take patient's diagnoses much earlier and treat them sooner and will be helpful in life-threatening situations. This causes healthcare to become more precise and responsive.

b. Inventory management

Hospitals can use optimized inventory management for their overall inventory and pharmacy. This will help healthcare technologies to track the shelf life of medicines, manage, and alert stock and alert the system for a low stock of medicines and

Challenges in Industrial Internet of Things (IIoT) 31

medical equipment and plan inventory ordering. This proper allocation of resources can effectively cut down the operating cost of the hospital.

2.2.4.2 Challenges of IIoT in Healthcare

The healthcare devices and sensors collect and transmit real-time data of the patient to the concerned authorities which need to be analyzed, process, and extract useful information. These generated huge data will likely need more storage requirements from terabytes to petabytes. This process of obtaining data, processing, analyzing, and storing in real-time from the vast number of devices will need vast infrastructure and skilled experts. AI-driven algorithms such as machine learning and deep learning, data analyst, and cloud are needed to organize these large data and to generate useful sense from them. This will need a lot of time and effort and will need time to mature. Also, the data collected from these devices should have common security practices or standards. These data contain personal and very delicate information of the patient, which can be used for unauthorized use such as blackmailing and extortion from the patients. The entire hospital network can be hacked and can affect smart health devices like heart-rate monitors, blood pressure readers, brain scanners, and can disrupt them. Also, with slow-changing regulations about medical equipment and heavy bureaucracy, the adoption of IIoT in the healthcare industry still has many challenges to be adopted fully.

2.3 CHALLENGES BASED ON IIOT COMPONENTS

The development of low-power processors, intelligent wireless networks, and a low-power sensor coupled with big data analytics is the key factor in the popularity of the IIoT. IIoT has three components or stages. Machines or devices embedded with sensors that collect data from the external environment, the network through which those data are transmitted, and the data from where useful insights are generated through different data analytics techniques. Hence, we explain challenges on IIoT based on these three different factors, as shown in Figure 2.2.

2.3.1 CHALLENGES IN DEVICES

The capture of data from the machines will be pivotal for making IIoT system effective. For that different sensor technology or devices can be employed for providing large data that can provide valuable insights. Information like the temperature, ambient light, and moisture of the industrial process can give the information about the required conditions for the machines to run correctly. Passive infrared devices are used for motion detection and RFID and GPS are used as tracking devices. In healthcare, sensor technology such as blood pressure monitor system, sugar level, etc., can give real-time information of patients that can give real-time information of patients and can be used for treatment to improve the condition of the patients. Devices such as industrial Robots, Smart Meters, smart beacon, and electronic shelf label have made automation in the industry possible. Since multiple sensors and devices are the key components of IIoT, it is also facing some challenges that could determine the success of the process.

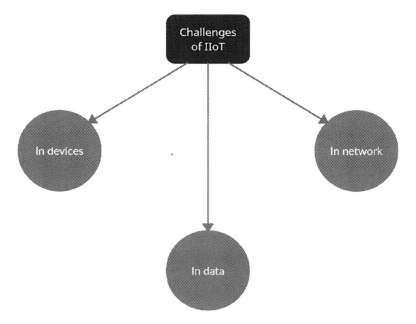

FIGURE 2.2 Challenges of IIoT based on three components.

With the development of digital smart devices, the increase in the number of smart devices used in daily life and the industrial environment has been improved exponentially and the number is in billions. These devices and sensors that can or have connected to the Internet have produced large streaming data. To be able to answer these real-time queries by this large number of IoT devices has been a challenge in IIoT.

Other important factors in IIoT applications are the precise placement of the sensors. If the network connection is wired, it will have a limited range. The wireless technology will provide a higher range but the need to power the wireless node by plugging in every few hours will be impractical and costly. The sensor deployed remotely will mostly be battery-powered objects. It will be uneconomical to change the batteries of the sensors deployed in fields regularly. Hence, to determine the control point of the sensors and manage the power consumption of the devices has been a challenge is it is critical in industrial IoT.

2.3.2 Challenges in Network

With the IIoT going more deeply toward the fourth industrial revolution, the requirement in the networks, communication, and information technologies has been changing. Most of the networks should be able to handle the requirements based on real-time, fast, and more reliable connections. The growth in the technological progress and performance of the network has been the backbone for the development of IIoT in general. Different industrial technology networks have been developed and used. The use of wired or optical fiber will have limited coverage of distance,

Challenges in Industrial Internet of Things (IIoT)

whereas 3G/4G network solutions are not as cheap as other network solutions. The range of wireless communication technology is available based on connectivity range, power requirements, and data transfer rates. Wired technology such as CAN bus, Ethernet, Modbus, Foundation Fieldbus, and PLC has already been established in industrial networks. In wireless protocols, the connections such as cellular-based networks that cover wide area networks such as LTE-M and 5G or short-range power-efficient one like Wi-Fi, LoRa, ZigBee, Z-wave, and BLE has been in use or still in progress. Among them, some support IIoT, while some are not compatible and outdated and will not meet the current communication requirement of IIoT. For example, the field buses network, which does not support the Ipv6, will not support IIoT now and it will not be able to gain the performance of high data transfer rate the IIoT demands. Other communication networks such as real-time Ethernet networks (RTE), which uses RVP/IP connectivity, wireless communication, and, more recently with the development of 5G technology [20], cellular networks have been proved suitable for IIoT technology. All these network systems have their standards and commercial protocols. A white paper has been published by the 5G alliance for connected industries which includes automation industries, telecom operators, and telecom vendors that address use cases and requirement of implementation of 5G over industrial application parameters that show the efficiency of these networks depends on its transmission speed, frame overhead, cycle time, and its stack processing time. These networks will play a significant part to make the adoption of IIoT more successful and to make them work in the automation of the industry. Hence, it also faces many challenges.

Industrial networks need to run for many years continuously as the visibility to the network is a key factor. Although how much powerful and vigorous network is developed, problems can still occur. Unexpected going down of Internet connectivity or even the new work going down for maintenance can cause data loss. It will be hard to guarantee zero-loss data during these connectivity issues in an industrial network. Also, the quality of the network can be affected by various environmental factors like variable critical temperature, mechanical stress, and interference during its operational life so it should include the ability to self-diagnose the problem, self-heal with the appropriate solution, and if not alert users and operators if issues arise are serious. This development of an efficient smart network system is still a challenge that needs to be solved.

Another challenge it faces is network security. Any IIoT devices that are controlled by network communication can be hacked. Software downloads or simple updates can attract or upload all malicious malware that can disrupt the network. Examples like BlackEnergy, Stuxnet, or German Steel malware mill can start a kinety cyberattack that can cause physical damage to the plant. The network should incorporate security measures and have safety regulations.

Depending on factors like coverage area, power consumption, security, and reliability, different industrial applications will have different requirements. Different network technology can have different standards. The performance of the networks depends mostly on the behavior of the channel, but only a few studies [21, 22] on this based on industry have been done. Although several key technologies are part of wireless networking standards such as WirelessHart (IEC62591) and emerging

34 Industrial Internet of Things

6TiSCH IETF standards for reliable and secure wireless sensor networks, still new global network standards for new technologies such as 5G are lagging for industrial applications.

2.3.3 CHALLENGES IN DATA

The key factors of IIoT are to accumulate real-time data from sensors, actuators, and change it into knowledge. The data produced by the IIoT system might be raw or pre-analyzed by the service provider, and either structured or unstructured. For this, the system that interconnects this with the sensor and controls across the large-scale field area network must be reliable. This end-to-end communication can provide valuable data from the sensor to its destination. The cloud will be the foundation for the storage and processing of data in cloud-based IoT. Not only the sensor like temperature, low pressure, and humidity are the source of data in IIoT, there are other sources of data that can be used by engineers and data scientists for smarter decision-making and faster responses across the organization. The data from supervisory control and data acquisition (SCADA) and enterprise assess management (EAM) system can be used for predictive maintenance, and from customer relationship management (CRM) can provide information to improve customer experience. Wearables like SmartCap that measures truck driver fatigue levels by monitoring their brain activity can give data that can be used for predicting where and when in the route they are likely to get fatigued. Data coming from the web like social media can be used for improving customer service. Location data coming from mobile devices, location beacons, and GIS systems can be used to optimize delivery routes in real time and track stolen or misplaced goods or containers. Machine-learning techniques can be used to predict certain events or accident before it happens and helps to put that data to good use for the safety of product and people. With these large data from IIoT being stored digitally, the challenges it faces in terms of data integrity, reliability, availability, scalability, robustness, and security [23] have been increased drastically.

With so many devices connect to the IIoT system, an enormous amount of data are generated. One of the main advantages of IIoT-based industry is the availability of large data but the same can be a challenge as it will need big infrastructure to store it. Also transforming these huge, complex, and unstructured data into useful data will need some sort of skills of a data analyst. Although some data lakes and cloud-based platforms can be used for this offline storage and integrated processing tools, managing these data in them also requires data management tools and platforms.

The data collected from a different source in the cloud may be used for different purposes such as marketing analysis. The data from industries such as industrial processes and medical records should have strict control procedures and authorization for their security from hacking or industrial espionage. These critical data should be used with the proper authorization of the data owner. They should be making assure of data leakage and piracy. For this, a proper security service should be developed in terms of data integrity with IIoT systems.

One of the key features of implementing IIoT in the industry is to use real-time data, gathered from different sources, in several frequencies and huge volumes to predict defects before they occur so that necessary action could be taken to prevent

Challenges in Industrial Internet of Things (IIoT)   35

them from happening. The real value of IIoT is using this unprecedented amount of quality data to think outside the box and drive business value which is what we are still learning. Data visualization immediately is a challenging task which also depends on the quality of data and their usefulness. So, it is challenging to prepare useful data from big raw data coming out from multiple connected devices and use them to make the organization more profitable in the IIoT systems. With more and more industries implementing IIoT in their system talent shortage for these data, analysis can be another challenge in the future. Also, the cost for the data acquisition, aggregation, and analysis will increase the investment of the industry.

2.4 FUTURE TECHNOLOGY AND ITS CHALLENGES

2.4.1 5G-BASED IIoT

5G has come after the three important stages of mobile communication technology 2G (voice digitizing), 3G (multimedia), and 4G (wireless broadband, respectively). 5G is to extend the advantage of mobile technology in new application fields, connect, control, exchange, locate, work together everything in an optimal way, and exceed the limits of space and time to create a new business model [24]. As implementing IIoT produces a huge amount of data transmission, processing of these massive data in real time can be achieved by a 5G wireless transmission network which has robustness, high transmission rate, high coverage, low latency, high reliability, and the massive number of devices [25]. As requirements of Industrial IoT applications are also similar, i.e., High security, system interoperability and integrity, effective deployment and engineering, and proactive maintenance emerge [26]. To solve the problem of large data transmission between M2M or between two peer nodes (D2D) in the network, 5G wireless technology network can achieve this with its antenna array, forming multi-beamforming directional transmission and loosely couple control mode. The data collected from manufacturing, assembly and logistics can be analyzed for real-time decision-making. Also, the security framework of the 5G wireless communication network which is based on a tamper resident Sim card can provide security regarding data privacy and security. High-speed data upload and feedback of calculation can be transmitted using 5G mobile wireless communication technology through small cell evolved node base of the network edge. The large-scale equipment interconnection and end-to-end transmission will be possible through the 5G network. Network latency can be reduced, as part of the data can be calculated and processed by the edge of the network to reduce the occupancy rate. Also, multi-beamforming directional transmission technology of 5G can lead to low energy consumption and low cost even at a large number of communication nodes [27].

Since 5G wireless communication technology is still in the development phase, there are many challenges to the implementation of 5G-based IIoT. Interference, attenuation of millimeter-wave can result in signal distortion while transmitting. Dynamic networking and layout optimization, and smart interaction of network collaborative control method are some of the challenges for future research in 5G-based IIoT. Security regarding confidentiality, integrity, availability, trustworthiness, and privacy have become key factors in the implementation of 5G-based IIoT. Being wireless technology 5G network has its cellular security lags and issues. Issues in wireless networks related to DoS attacks like jamming can be avoided by using spread-spectrum techniques in the cellular network but cannot be applied in IoT nodes with limited resources. These types of specific problems still need to be studied to find concrete solutions. Also, due to different implementation approached in multiple countries such as Russia, China, South Korea, and the European Union which have outlined their definition of 5G networks, speed, and regulations outlining where 5G may occur, it will be hard for 5G wireless standard to be global. Hence, coexistence of different wireless protocols and systems will be a big challenge for adopting 5G-based IIoT [28].

2.4.2 BLOCKCHAIN-BASED IIoT

Blockchain is a real-time shared and immutable ledger of records that are stored in a distributed point-to-point manner and has no central authority. Users have the right to access and edit the blocks for which they have the private key. It uses motion time stamping, distributed consensus, data encryption, and economic incentives [29]. Most of the data collected from the connected devices in IIoT share those data with the cloud. These data can be easily attacked by cyber-terrorist to expand their botnets and can be used to exploit vulnerabilities. The blockchain has been an alternative to solve this IIoT security issues. Many research [30, 31] has proposed blockchain-based IIoT. A localized blockchain-enabled secure energy trading system among vehicles was built [32] for the demand of increasing resource requirements due to the increasing number of devices. The blockchain also supports direct device to device data exchange and sharing for mobile commerce without any third-party involvement. First, a smart contract is built between the seller and buyer with exact transaction information. After both parties confirm, it is publishing in a blockchain system and has secure connections. Other features of blockchain which are time-stamped make it possible to traced back to a specific time to ensure the security of the supply chain. The product can be traced to the source of the raw materials and the transaction can give information about the faulty IIoT device if the product recalls because of security breaches. This has found time saving and efficient as an example shows that with blockchain it just need 2.2 s to get detailed information about goods which otherwise used to take around 16 days 18 hr and 26 min [33]. Blockchain identity authentication and access management techniques can enhance IIoT security. It has been used in different application field to prevent hacking and data breaches such as in some healthcare area where company uses blockchain to record patient's information.

Although the use of blockchain in IIoT has shown large possibility, it is still in the experimental stage. More research and study should be done to know about the efficiency of this in the different application areas. One of the great challenges that it faces is to know how it will perform with the increase in the number of transactions. With an increase in the number of machines, devices, and data in the network, the transactions increase, and the complexity of the mining algorithms also increases. The computing and data exchange required by blockchains need high energy and

Challenges in Industrial Internet of Things (IIoT) 37

computing power which may not be suitable for low-power portable devices. So, designing energy-efficient blockchain protocols and algorithms is one of the great challenges of blockchain-based IIoT.

Also, there is a lack of responsive international standards for using blockchain-based IIoT. There are no such criteria to test the stability, performance, and security of the application. Motivating the industry to develop blockchain-based IIoT in their company without the valid verification standards will be another challenge for blockchain-based IIoT.

2.5 CONCLUSION

With the extensive application of IoT in the industry, the implementation in the industry has been effective to enhance production efficiency, supply chain, and real-time assets monitoring. The adoption of IIoT in the industry brings interaction of complex hardware, software, big data, network components, and the demanding requirement of security, safety, and privacy of the data and system. With more smart devices connecting to the Internet, generating more useful big data, the use of Artificial intelligence can bring insight information from the data that can be used for predictive analysis for increasing production and reducing errors and disruptions.

The chapter describes some of the challenges existing in IIoT based on four main application fields like the manufacturing industry, agricultural industry, logistic industry, and healthcare based on specific areas of those fields and in three different factors of IIoT: data, device, and network. There are still many challenges and issues that need to be addressed for IIoT to be fully applicable. The most common challenges include proper management of data, security, safety, technical collaboration and requirements, lack of skilled manpower, and standards and protocols. As IIoT is still in the early stage of its applications, proper study and research should overcome these challenges. Furthermore, new technologies adoption in IIoT along with its possible challenges are described.

BIBLIOGRAPHY

[1] L. D. Xu, W. He, and S. Li, "Internet of things in industries: a survey," *IEEE Trans. Ind. Informatics*, vol. 10, no. 4, pp. 2233–2243, 2014.

[2] S. K. Rao and R. Prasad, "Impact of 5G technologies on industry 4.0," *Wirel. Pers. Commun.*, vol. 100, no. 1, pp. 145–159, 2018.

[3] "Industry 4.0: a complete guide," *Stefanini.com*, 2019. [Online]. Available: https://stefanini.com/en/trends/news/industry-4-a-complete-guide.

[4] F. Tao, J. Cheng, and Q. Qi, "IIHub: an industrial Internet-of-Things hub toward smart manufacturing based on cyber-physical system," *IEEE Trans. Ind. Informatics*, vol. 14, no. 5, pp. 2271–2280, 2017.

[5] F. Tao, J. Cheng, Q. Qi, M. Zhang, H. Zhang, and F. Sui, "Digital twin-driven product design, manufacturing and service with big data," *Int. J. Adv. Manuf. Technol.*, vol. 94, no. 9, pp. 3563–3576, 2018.

[6] J. Biron, "The State of the Industrial Internet of Things 2017."

[7] "Industrial Iot Market | Meticulous Market Research Pvt. Ltd.," 2020. [Online]. Available: https://www.meticulousresearch.com/product/industrial-iot-market-5102. [Accessed: 31-Jan-2021].

[8] A. Gilchrist, Introducing industry 4.0. In: *Industry 4.0*. Berkeley, CA: Apress, 2016. doi:10.1007/978-1-4842-2047-4_13

[9] Y. Zhang, Z. Guo, J. Lv, and Y. Liu, "A framework for smart production-logistics systems based on cps and industrial IoT," *IEEE Trans. Ind. Informatics*, vol. 14, no. 9, pp. 4019–4032, 2018.

[10] D. Vasisht, Z. Kapetanovic, J. Won, X. Jin, R. Chandra, S. Sinha, A. Kapoor, M. Sudarshan, and S. Stratman, *"Farmbeats: an IoT platform for data-driven agriculture,"* In: *14th {USENIX} Symposium on Networked Systems Design and Implementation ({NSDI} 17)*, pp. 515–529, Boston, USA, 2017.

[11] "FarmBeats: AI, Edge & IoT for Agriculture - Microsoft Research," 2020. [Online]. Available: https://www.microsoft.com/en-us/research/project/farmbeats-iot-agriculture/. [Accessed: 30-Jan-2021].

[12] "Digital Farming decisions and insights to maximize every acre," 2020. [Online]. Available: https://www.climate.com/. [Accessed: 30-Jan-2021]

[13] S. Kinney, "World Economic Forum: tech takes from The Davos Agenda," *RCR Wireless News*, 2021. [Online]. Available: https://www.rcrwireless.com/20210129/policy/world-economic-forum-tech-takes-from-the-davos-agenda

[14] "Digital Transformation Monitor Industry 4.0 in agriculture: Focus on IoT aspects," 2017. [Online]. Available: https://ec.europa.eu/growth/tools-databases/dem/monitor/sites/default/files/DTM_Agriculture 4.0 IoT v1.pdf.

[15] "CILT(UK) > Knowledge > Knowledge Bank > Resources > Other Resources > Useful glossaries," 2020.

[16] A. Abdelhadi and E. Akkartal, "A framework of IoT implementations and challenges in warehouse management, transportation and retailing," *Eurasian Bus. & Econ. J.*, vol. 18, pp. 25–41, 2019.

[17] B. Farahani, F. Firouzi, V. Chang, M. Badaroglu, N. Constant, and K. Mankodiya, "Towards fog-driven IoT eHealth: promises and challenges of IoT in medicine and healthcare," *Futur. Gener. Comput. Syst.*, vol. 78, pp. 659–676, 2018.

[18] T. J. McCue, "$117 billion market for internet of things in healthcare by 2020," *Forbes Tech*, 2015. [Online]. Available: https://www.forbes.com/sites/tjmccue/2015/04/22/117-billion-market-for-internet-of-things-in-healthcare-by-2020/?sh=7a69aa1b69d9

[19] M. S. Hossain and G. Muhammad, "Cloud-assisted Industrial Internet of Things (IIoT) – Enabled framework for health monitoring," *Comput. Networks*, vol. 101, pp. 192–202, 2016.

[20] S. Mumtaz, A. Bo, A. Al-Dulaimi, and K. Tsang, "Guest editorial 5G and beyond mobile technologies and applications for Industrial IoT (IIoT)," *IEEE Trans. Ind. Informatics*, vol. 14, no. 6, pp. 2588–2591, 2018.

[21] A. Willig, M. Kubisch, C. Hoene, and A. Wolisz, "Measurements of a wireless link in an industrial environment using an IEEE 802.11-compliant physical layer," *IEEE Trans. Ind. Electron.*, vol. 49, no. 6, pp. 1265–1282, 2002.

[22] E. Tanghe et al., "The industrial indoor channel: large-scale and temporal fading at 900, 2400, and 5200 MHz," *IEEE Trans. Wirel. Commun.*, vol. 7, no. 7, pp. 2740–2751, Jul. 2008.

[23] D. Mourtzis, E. Vlachou, and N. Milas, "Industrial big data as a result of IoT adoption in manufacturing," *Procedia CIRP*, vol. 55, pp. 290–295, 2016.

[24] M. Agiwal, A. Roy, and N. Saxena, "Next generation 5G wireless networks: a comprehensive survey," *IEEE Commun. Surv. Tutorials*, vol. 18, no. 3, pp. 1617–1655, 2016.

[25] F. Voigtländer et al., "5G for the Factory of the Future: Wireless Communication in an Industrial Environment," *arXiv Prepr. arXiv1904.01476*, 2019.

[26] D. Kozma, P. Varga, and G. Soós, *"Supporting digital production, product lifecycle and supply chain management in industry 4.0 by the arrowhead framework–a survey,"* In: *2019 IEEE 17th International Conference on Industrial Informatics (INDIN)*, Vol. 1, pp. 126–131, IEEE, 2019, July.

[27] J. Cheng, W. Chen, F. Tao, and C.-L. Lin, "Industrial IoT in 5G environment towards smart manufacturing," *J. Ind. Inf. Integr.*, vol. 10, pp. 10–19, 2018.

[28] E. O'Connell, D. Moore, and T. Newe, "Challenges associated with implementing 5G in manufacturing," *Telecom*, vol. 1, no. 1, pp. 48–67, 2020.

[29] Q. Wang, X. Zhu, Y. Ni, L. Gu, and H. Zhu, "Blockchain for the IoT and industrial IoT: a review," *IoT*, vol. 10, p. 100081, 2020.

[30] N. Teslya and I. Ryabchikov, *"Blockchain-based platform architecture for industrial IoT,"* In: *2017 21st Conference of Open Innovations Association (FRUCT)*, pp. 321–329, Helsinki, Finland, IEEE, 2017, November.

[31] A. Bahga and V. K. Madisetti, "Blockchain platform for industrial internet of things," *J. Softw. Eng. Appl.*, vol. 9, no. 10, pp. 533–546, 2016.

[32] J. Kang, R. Yu, X. Huang, S. Maharjan, Y. Zhang, and E. Hossain, "Enabling localized peer-to-peer electricity trading among plug-in hybrid electric vehicles using consortium blockchains," *IEEE Trans. Ind. Informatics*, vol. 13, no. 6, pp. 3154–3164, 2017.

[33] S. Charlebois, "How blockchain technology could transform the food industry," *The Conversation* 20, 2017. [Online]. Available: https://theconversation.com/how-blockchain-technology-could-transform-the-food-industry-89348

3 IoT-Based Automated Healthcare System

Darpan Anand and Aashish Kumar

Chandigarh University, Chandigarh, India

CONTENTS

3.1 Introduction 41
    3.1.1 Software-Defined Network 42
    3.1.2 Network Function Virtualization...43
    3.1.3 Sensor Used in IoT Devices...43
3.2 SDN-Based IoT Framework...47
3.3 Literature Survey...49
3.4 Architecture of SDN-based IoT for Healthcare System 50
3.5 Challenges 51
3.6 Conclusion...51
References..52

3.1 INTRODUCTION

In many countries, health is the primary concern that affects the quality of life. Shaik and Chitre [1] demonstrated that the non-presence of electronic medical services frameworks and over-reliance on paper-based frameworks in numerous essential medical care centers in rustic regions prompted patients to keep their clinical records without anyone else. In a few examples, this has prompted a patient's passing away because of the inaccessibility of a doctor. Additionally, medical attendants in clinics gather indispensable sign information, for example, circulatory strain, temperature, respiratory rate, and beat rate to screen patient advancement and abnormalities [2]. The attendants commonly record these essential sign readings of patients physically, which is manageable to numerous blunders in the data [3]. According to Vikash et al. [4], the worldwide medical services will increment at an accumulated yearly development rate of 5.6% and will arrive at 25 billion by 2020. Managing billions of devices not easy. This leads to the problem of network complexity, network delay, vendor-specific components, etc. Software-defined network (SDN) and network function virtualization solved these issues. SDN-based Internet of Things (IoT) devices have a large number of advantages such as:

1. Network management is easy
2. Security and privacy is enhanced
3. Accessing information from anywhere

DOI: 10.1201/9781003102267-3

4. Efficient use of resource utilization
5. Energy management

These SDN-based IoT devices have a large effect on healthcare. SDN-based IoT architecture is divided into several parts: namely, the data plane layer, core backbone network, SDN controller, and datacenter network. Apart from benefits, there are a lot of challenges that SDN-based IoT devices have to face in the healthcare system.

3.1.1 Software-Defined Network

In a traditional network, the network plane and control plane are tightly coupled; this led to the problem of dynamic IP assignment, change in steering, data transfer capacity the executives start to finish reachability, etc. To solve these problems, the SDN came into the picture. SDN [5] separates the control plane and information plane. It is currently used in various data centers of big giant companies, such as Amazon, Google, Facebook, etc., and in a 5g network [6]. The SDN architecture is shown in Figure 3.1 [7–9].

Infrastructure layer: These are the dumb switches and routers, which route the packets according to the set of instructions defined in the forwarding table [10].
Control layer: It is called "brain of the network". The controller has the global view of the network, so that management and securing the network become easy as compared with traditional network [11].
Management plane: It is used to access and helps in the management of our network devices [12].

The correspondence between the foundation layer and control layer is performed by using the OpenFlow protocol [13]. It is the most commonly utilized southward

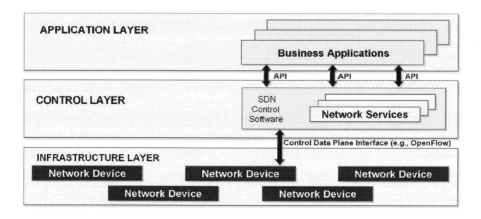

FIGURE 3.1 SDN architecture.

IoT-Based Automated Healthcare System

API in the SDN and created, normalized by the Open Flow Networking Foundation (ONF). SDN integrated with IoT devices [14] has many advantages that are discussed in the latter part of this section.

3.1.2 NETWORK FUNCTION VIRTUALIZATION

A conventional organization depends on the idea of "One Node One Service", which implies that an assistance is sent on explicit equipment. This conventional organization causes versatility, adaptability, speed decrease, delay, and other organization dormancy issues. These issues will be harmful to the business, the association, etc. Consequently, we need a stage to overcome the previously mentioned issues and it is reasonable to determine network function virtualization (NFV) issues with virtualization of organization assets. NFV utilizes three systems as virtualization [15], programming [16], and orchestration [17] for network virtualization, as shown in Figure 3.2. NFV is likewise helpful for supplanting costly, devoted, and explicit reason equipment with nonexclusive equipment.

Now, we will discuss about the sensors used in IoT devices in detail.

3.1.3 SENSOR USED IN IoT DEVICES

1. Ring sensor: It measures the human heart rate and oxygen concentration. It is also called as pulse oximetry sensor [18]. (Figures 3.3 to 3.10)
2. ECG sensor: It measures the electrical and macular function of the heart. It continuously measures the heart rate of the human body [19].
3. GSR sensor: GSR stands for galvanic skin response sensor. It can be used to detect the emotion and stress of the human body [20].
4. Graphene vapor sensor: It measures chemical evaporation through skin. It can also detect diabetes, high blood pressure, and anemia [21].
5. Health patch sensor: It detects chronic disease and can be fixed with the chest of the human body. It measures the temperature, heart rate, and ECG of the human body [22].
6. Oximeter sensor: It detects the amount of oxygen present in the blood (hemoglobin)[23].
7. Q^{TM} sensor: It is a sensor that detects the emotion of the human by the response of physiological behavior [24].
8. Airflow sensor. It is a thin nasal sensor that detects the respiratory rate of the patients by using prongs. It detects asthma, anxiety, pneumonia, and drug overdose [25].
9. Optical biosensor: It detects the continuous signs from the optical creation of biomolecules [26]
10. Respiration sensor: It measures the rate of thoracic breathing as well as normal breathing [27].

Till now, we have discussed an overview of SDN, NFV, and sensors used in IoT devices. Section 2 discusses about SDN-based IoT framework.

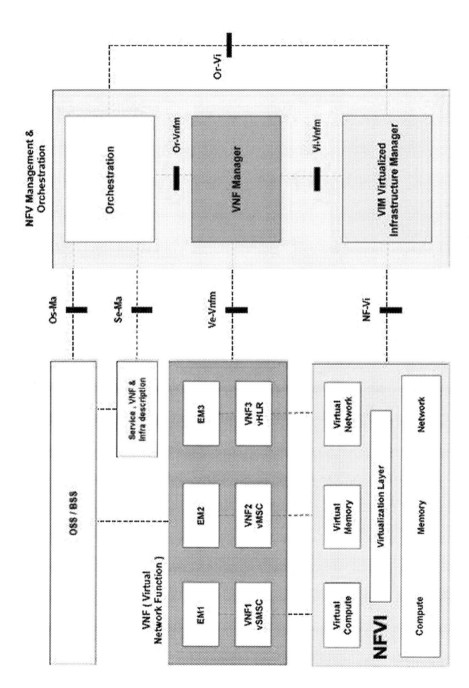

FIGURE 3.2 NFV architecture.

IoT-Based Automated Healthcare System 45

FIGURE 3.3 Ring sensor.

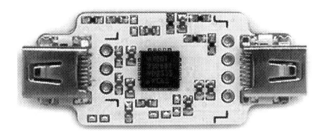

FIGURE 3.4 ECG sensor.

FIGURE 3.5 GSR sensor.

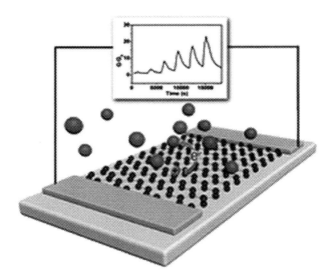

FIGURE 3.6 Graphene vapor sensor.

FIGURE 3.7 Health patch sensor.

FIGURE 3.8 Oximeter sensor.

IoT-Based Automated Healthcare System

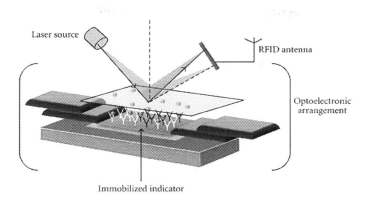

FIGURE 3.9 Optical biosensor.

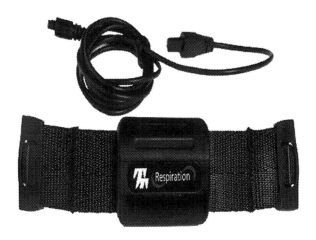

FIGURE 3.10 Respiration sensor.

3.2 SDN-BASED IoT FRAMEWORK

The following section discusses the advantages of the SDN-based IoT framework [28], which are as follows:

(1) **Network management**: In this era, millions of devices are interconnected by the IoT technology; therefore, a large amount of information is generated and needs to be executed efficiently. To solve the above problem, a separate technology is introduced, i.e. SDN. As we have discussed, SDN has a global view of the network and solves the problem of network delay, bandwidth utilization, etc.

Accessing information from anywhere: Since the billions of equipment are joined through IoT technology, the equipment can be accessed from anywhere around the world and functionality can be changed according to the user's need. This can be achieved by SDN technology.

(2) **Resource utilization:** Overexploitation of resources or underutilization of resources degrades the network performance. To achieve high network performance, the overall view of the entire network should be available. SDN solves the problem.

(3) **Energy management:** IoT devices are battery-operated devices, and therefore are power-constraint devices. SDN utilizes the resources efficiently and reliably and led to the efficient utilization of power resources that take place.

Figure 3.11 shows software-defined IoT framework. The framework is divided into four parts.

(1) **Data plane layer:** It has IoT devices such as ECG machines, pulse oximeter, etc. These devices are connected to SDN switches and routers through Wi-Fi or any other LTE technology. Routers may be Juniper MX-Series, PC engines, Pronto 3220/3290, HP curve, etc. Their main aim is to collect the data from the patients.

(2) **Core backbone network:** This is Internet network through which the information is collected from sensing devices and transferred to the controller and data center network.

(3) **SDN controller:** It has the global view of the network; therefore, the network management and routing of data can be done efficiently. It can be located in various geographical areas.

(4) **Datacenter network**: In this network, actual processing of the sensed data takes place and can be used for further use.

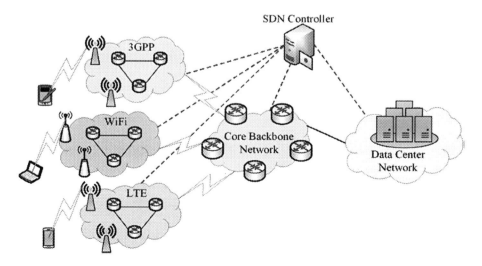

FIGURE 3.11 A schematic view of software-defined IoT.

IoT-Based Automated Healthcare System

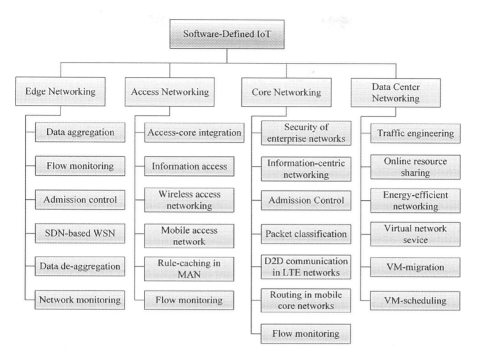

FIGURE 3.12 Overview of different aspects of SDN-based IoT networks.

Figure 3.12 shows the SDN-based IoT networks.

3.3 LITERATURE SURVEY

Salahuddin et al. [16] have considered the test of gathering and totaling and playing out the investigation on the organization information from different IoT gadgets. This crude information can be huge information and can occur in any form such as sound, video, and text. For continuous transmission of the large measure of information, creators have proposed a unified and adaptable framework design that can give the security, protection, and other security fundamentals for different associated medical care applications.

Utilization of the IoT, its experience, and benefits are proposed in Hu et al. [29]. The subtleties of the planning are additionally represented by them. Creators have proposed engineering on IoT-medical services along with the further different examination headings and difficulties.

In Khayat et al. [30], researchers have analyzed the troubles and various problems related to weak points and assurance of the standard medical administrations monitoring system (HMS). This plan can be utilized to regulate IoT devices based on clinical benefits checking structures. It guarantees the security, protection as well as the dependability of the different conveyed benefits. These administrations and strategies are expected for old individuals and patients.

This chapter depicts the situation with IoT with various measurements [31]. The eight examination headings, specifically enormous scaling, designing and conditions, making data, force, responsiveness, security, assurance, protection, and human on top of it, have been examined to give an elaborated usage of sensors and actuators [31].

The chapter proposed a blend of advancements specifically SDN and IoT [32]. New improvements in different areas like remote and optical space are eatured to combine the innovation. SDN innovation is not only a structure with a focus on industry and the scholarly world, but also it is the best innovation in diminishing the network with a predefined convention in the innovative space.

SDN from different points of view tackles the issue of productivity, versatility, sensibility, and cost-adequacy in IoT [33]. The different system administration perspectives such as edge, access, and center have been also depicted. The various plans related to framework organization have furthermore been portrayed to support the association with SDN [33].

3.4 ARCHITECTURE OF SDN-BASED IoT FOR HEALTHCARE SYSTEM

Figure 3.13 outlines the architecture of SDN-based IoT medical services. The network is incorporated further into different associations of different crisis facilities. These facilities are moreover connected with different work where data exchange from various IoT contraptions is cultivated. It might be seen that network is incorporated further into different associations of different crisis facilities. These facilities are moreover connected with the subnetworks in which data exchange from various IoT contraptions, sufferer data, etc., are cultivated. The server farm at which all the connected information is put away is halfway positioned. The utility at server is additionally included a three-layered engineering. The principal layer is the system's administration layer in which the organization geographies, such as physical setup and organizing gadgets (switches, switch), are situated. The layer above the principal layer consists of SDN regulator. Conventional organizations include various system administration gadgets. These gadgets comprise the insight and the hidden

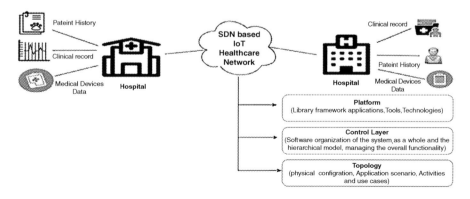

FIGURE 3.13 Architecture of SDN-based IoT for healthcare system.

IoT-Based Automated Healthcare System 51

organization framework inside it; however, SDN changes the entire thought. The different organizing gadgets utilized at the foundation layer as referenced before are only the idiotic gadgets that are utilized to convey the traffic from one spot to another. The guidance to them is given by the SDN regulator as it were. The end layer is the client layer where applications such as firewall, intrusion detection, etc., are installed. Using these applications in an SDN-based association improves the challenges.

3.5 CHALLENGES

Chen et al. [34] classify the difficulties in IoT into four sections: hardware challenges, architecture challenges, technical challenges, and security and security issues. In the plan part, the coordination, comprehension of data, and heterogeneous reference models have been analyzed, with the objective that flexibility and information on the device can be improved. In the specific test, issues related to arranging accessibility and distant correspondence with the reliable organization have been analyzed. In gear challenge, the issues related to control usage and cost-practicality have been portrayed. In insurance and security challenges, the instrument and concentrated game plans oversee low inactivity and flexibility issues in hardware; standard test discusses the coordination of IoT establishments with multiparty. The organized exertion of particular requirements with that of utilization circumstances has been tended to in a business challenge.

Čolaković and Hadžialić [35] introduced difficulties such as normalization, framework design, interoperability and reconciliation, accessibility and unwavering quality, information stockpiling and handling, adaptability, the board and self-arrangement, execution and QoS, ID and interesting personality, force and energy utilization, security, protection, and natural issues.

Lee et al. [36] portrayed the difficulties to order the spaces. The enrolling stage covers the blend of standard articles with the sensors having negligible exertion and low force usage. The idea of inertness has likewise been examined. The connection among calculation and correspondence to decrease the utilization of IoT gadgets is also clarified.

In Miorandi et al. [37], the vital difficulties in IoT and SDN are "naming", "characterizing the board", "interoperability and normalization", "data protection", "objects and security", "organization security", "information classification"and "encryption, range, greening" of IoT; and all these have been depicted in Sood et al. [32].

3.6 CONCLUSION

Khan et al. [38] explained about the imperative troubles in IoT and SDN. Some of them are naming and character the board, interoperability and standardization, information assurance, objects prosperity and security, association security, data grouping and encryption, range, greening of IoT. All these have also been portrayed in Sood et al. [32]. The makers have focused on the outstanding character of sensors, interoperability etc. This chapter explains the concept of the IoT for the healthcare sector. It is very important for this sector due to migration to the industry 4.0. There are various sensors available to measure various health parameters, such as health patch sensor, oximeter sensor, etc., to measure various important health parameters. The use of these technologies has been seen by all of us in the case of COVID-19 pandemic as well where the temperature, oxygen level, etc., are also measured through these devices. Furthermore, there are various devices and systems that need to be developed to identify vulnerable patients. It is also good for critical patients to monitor them based on the real-time health parameters. Therefore, the chapter is very important to know about the IoT concepts used in the sector of healthcare to make it under the flagship of the industry 4.0 i.e. healthcare 4.0.

REFERENCES

[1] Shaikh, S. & Chitre, V. (2017). *Healthcare monitoring system using IoT*. In: *2017 International Conference on Trends in Electronics and Informatics (ICEI)* (pp. 374–377).

[2] Evans, D., Hodgkinson, B., & Berry, J. (2001). Vital signs in hospital patients: a systematic review. *IJNS*, 38(6), 643–650.

[3] Yang, Z., Kankanhalli, A., Ng, B.-Y., & Lim, J. T. Y. (2015). Examining the pre-adoption stages of healthcare IT: a case study of vital signs monitoring systems. *Information & Management*, 52(4), 454–467.

[4] Vikash, M., Gupta, M., & Upadhayay, Kumar S. (2015). *A survey on wireless body area network: security technology and its design methodology issue*. In: *2015 International Conference on Innovations in Information, Embedded and Communication Systems (ICIIECS)* (pp. 328–337, Coimbatore, India.

[5] Xia, W., Wen, Y., Foh, C. H., Niyato, D., & Xie, H. (2014). A survey on software-defined networking. *IEEE Communications Surveys & Tutorials*, 17(1), 27–51.

[6] Gupta, A. & Jha, R. K. (2015). A survey of 5G network: Architecture and emerging technologies. *IEEE Access*, 3, 1206–1232. doi:10.1109/ACCESS.2015.2461602

[7] Hu, Z., Wang, M., Yan, X., Yin, Y., & Luo, Z. (2015, February). *A comprehensive security architecture for SDN*. In: *2015 18th International Conference on Intelligence in Next Generation Networks* (pp. 30–37). Paris, France, IEEE. doi:10.1109/ICIN.2015.7073803

[8] Hayes, M., Ng, B., Pekar, A., & Seah, W. K. (2017). Scalable architecture for SDN traffic classification. *IEEE Systems Journal*, 12(4), 3203–3214. doi:10.1109/JSYST.2017.2690259

[9] Sufiev, H. & Haddad, Y. (2016, November). *A dynamic load balancing architecture for SDN*. In: *2016 IEEE International Conference on the Science of Electrical Engineering (ICSEE)* (pp. 1–3). Eilat, Israel, IEEE.

[10] Dhamecha, K. & Trivedi, B. (2013). SDN issues – a survey. *International Journal of Computer Applications*, 73(18).

[11] Jimenez, Y., Cervello-Pastor, C., & Garcia, A. J. (2014, June). *On the controller placement for designing a distributed SDN control layer*. In: *2014 IFIP Networking Conference* (pp. 1–9). IEEE.

[12] Wang, Y., & Matta, I. (2014, October). *SDN management layer: design requirements and future direction*. In: *2014 IEEE 22nd International Conference on Network Protocols* (pp. 555–562). IEEE.

[13] McKeown, N., Anderson, T., Balakrishnan, H., Parulkar, G., Peterson, L., Rexford, J., Shenker, S., & Turner, J. (2008). OpenFlow: enabling innovation in campus networks. *ACM SIGCOMM Computer Communication Review*, 38(2), 69–74.

[14] Huh, S., Cho, S., & Kim, S. (2017, February). *Managing IoT devices using blockchain platform*. In: *2017 19th International Conference on Advanced Communication Technology (ICACT)* (pp. 464–467). PyeongChang, Korea (South), IEEE. doi:10.23919/ICACT.2017.7890132

IoT-Based Automated Healthcare System

[15] Chiueh, S. N. T. C. & Brook, S. (2005). A survey on virtualization technologies. *Rpe Report*, 142.

[16] Salahuddin, M. A., Al-Fuqaha, A., Guizani, M., Shuaib, K., & Sallabi, F. (2018). Softwarization of internet of things infrastructure for secure and smart healthcare. arXiv preprint arXiv:1805.11011.

[17] Adler, S. & Hesterman, P. (1989). *The study of orchestration* (Vol. 4, p. 1024). New York, NY: WW Norton. ISBN-13: 978-0393920659

[18] Rhee, S., Yang, B. H., Chang, K., & Asada, H. H. (1998, October). *The ring sensor: a new ambulatory wearable sensor for twenty-four hour patient monitoring*. In: *Proceedings of the 20th Annual International Conference of the IEEE Engineering in Medicine and Biology Society* (Vol. 20, No. 4, pp. 1906–1909).

[19] Nemati, E., Deen, M. J., & Mondal, T. (2012). A wireless wearable ECG sensor for long-term applications. *IEEE Communications Magazine*, 50(1), 36–43.

[20] Bakker, J., Pechenizkiy, M., & Sidorova, N. (2011, December). *What's your current stress level? Detection of stress patterns from GSR sensor data*. In: *2011 IEEE 11th International Conference on Data Mining Workshops* (pp. 573–580). IEEE.

[21] Bogue, R. (2014). Graphene sensors: a review of recent developments. *Sensor Review*.

[22] Wu, T., Wu, F., Qiu, C., Redouté, J. M., & Yuce, M. R. (2020). A rigid-flex wearable health monitoring sensor patch for IoT-connected healthcare applications. *IEEE Internet of Things Journal*, 7(8), 6932–6945.

[23] Fernandez, M., Burns, K., Calhoun, B., George, S., Martin, B., & Weaver, C. (2007). Evaluation of a new pulse oximeter sensor. *American Journal of Critical Care*, 16(2), 146–152.

[24] Zareie, S., Khosravi, H., Nasiri, A., & Dastorani, M. (2016). Using Landsat Thematic Mapper (TM) sensor to detect change in land surface temperature in relation to land use change in Yazd, Iran. *Solid Earth*, 7(6), 1551–1564.

[25] Zhao, Y., Wang, P., Lv, R., & Liu, X. (2016). Highly sensitive airflow sensor based on Fabry–Perot interferometer and Vernier effect. *Journal of Lightwave Technology*, 34(23), 5351–5356.

[26] Cross, G. H., Reeves, A. A., Brand, S., Popplewell, J. F., Peel, L. L., Swann, M. J., & Freeman, N. J. (2003). A new quantitative optical biosensor for protein characterisation. *Biosensors and Bioelectronics*, 19(4), 383–390.

[27] Güder, F., Ainla, A., Redston, J., Mosadegh, B., Glavan, A., Martin, T. J., & Whitesides, G. M. (2016). Paper-based electrical respiration sensor. *Angewandte Chemie International Edition*, 55(19), 5727–5732. Wiley. doi:10.1002/anie.201511805

[28] Li, J., Altman, E., & Touati, C. (2015). A general SDN-based IoT framework with NVF implementation. *ZTE Communications*, 13(3), 42–45.

[29] Hu, L., Qiu, M., Song, J., Hossain, M. S., & Ghoneim, A. (2015). Software defined healthcare networks. *IEEE Wireless Communications*, 22(6), 67–75.

[30] Khayat, M., Barka, E., & Sallabi, F. (2019, July). *SDN_Based SecureHealthcare Monitoring System (SDN-SHMS)*. In: *2019 28th International Conference on Computer Communication and Networks (ICCCN)* (pp. 1–7). IEEE.

[31] Stankovic, J. A. (2014). Research directions for the internet of things. *IEEE Internet of Things Journal*, 1(1), 3–9.

[32] Sood, K., Yu, S., & Xiang, Y. (2015). Software-defined wireless networking opportunities and challenges for Internet-of-Things: a review. *IEEE Internet of Things Journal*, 3(4), 453–463.

[33] Badotra, S. & Panda, S. N. (2019). A review on software-defined networking enabled Iot cloud computing. *IIUM Engineering Journal*, 20(2), 105–126.

[34] Chen, S., Xu, H., Liu, D., Hu, B., & Wang, H. (2014). A vision of IoT: applications, challenges, and opportunities with china perspective. *IEEE Internet of Things Journal*, 1(4), 349–359. IEEE. doi:10.1109/JIOT.2014.2337336

[35] Čolaković, A., & Hadžialić, M. (2018). Internet of Things (IoT): A review of enabling technologies, challenges, and open research issues Computer Networks, *Computer Networks*, 144, 17–39.

[36] Lee, S., Bae, M., & Kim, H. (2017). Future of IoT networks: a survey. *Applied Sciences*, 7(10), 1072.

[37] Miorandi, D., Sicari, S., De Pellegrini, F., & Chlamtac, I. (2012). Internet of things: vision, applications and research challenges. *Ad hoc Networks*, 10(7), 1497–1516. Elsevier. doi:10.1016/j.adhoc.2012.02.016

[38] Khan, R., Khan, S. U., Zaheer, R., & Khan, S. (2012, December). *Future internet: the internet of things architecture, possible applications and key challenges.* In: *2012 10th international conference on frontiers of information technology* (pp. 257–260). IEEE.

[39] Hawilo, H., Shami, A., Mirahmadi, M., & Asal, R. (2014). NFV: state of the art, challenges, and implementation in next generation mobile networks (vEPC). *IEEE Network*, 28(6), 18–26.

[40] Kim, H., & Feamster, N. (2013). Improving network management with software defined networking. *IEEE Communications Magazine*, 51(2), 114–119.

[41] Hamed, M. I., ElHalawany, B. M., Fouda, M. M., & Eldien, A. S. T. (2017, December). *A novel approach for resource utilization and management in SDN.* In: *2017 13th International Computer Engineering Conference (ICENCO)* (pp. 337–342). IEEE.

[42] Aujla, G. S., & Kumar, N. (2018). SDN-based energy management scheme for sustainability of data centers: An analysis on renewable energy sources and electric vehicles participation. *Journal of Parallel and Distributed Computing*, 117, 228–245.

[43] Zarca, A. M., Bernabe, J. B., Trapero, R., Rivera, D., Villalobos, J., Skarmeta, A., Bianchi, S., Zafeiropoulos, A., & Gouvas, P. (2019). Security management architecture for NFV/SDN-aware IoT systems. *IEEE Internet of Things Journal*, 6(5), 8005–8020.

4 Internet of Things (IoT)-Based Industrial Monitoring System

Syeda Florence Madina, Md. Shahinur Islam, and Fakir Mashque Alamgir

East West University, Dhaka, Bangladesh

Mohammad Farhan Ferdous

Japan-Bangladesh Robotics and Advance Technology
Research Centre (JBRATRC), Dhaka, Bangladesh

CONTENTS

4.1 Introduction 56
    4.1.1 Background 57
    4.1.2 Project Organization 57
4.2 Literature Review 57
    4.2.1 "IoT" and Its Smart Applications 57
    4.2.2 Health Monitoring and Management Using Internet-of-Things
        (IoT) Sensing with Cloud-based Processing: Opportunities and
        Challenges..58
    4.2.3 IoT (Internet of Things)-Based Monitoring and Control System
        for Home Automation 59
    4.2.4 Industrial Automation Using IoT 59
4.3 Hardware and Software 59
    4.3.1 Hardware 59
        4.3.1.1 NODEMCU...60
        4.3.1.2 ESP8266 60
        4.3.1.3 Gas Sensor...61
        4.3.1.4 DHT11..61
        4.3.1.5 PIR Sensor...62
        4.3.1.6 Breadboard 63
        4.3.1.7 Power Board 63
        4.3.1.8 MCP3008 63
        4.3.1.9 Buzzer 64
    4.3.2 Software 64
        4.3.2.1 Web Server 64

DOI: 10.1201/9781003102267-4

		4.3.2.2 Arduino Studio	65
4.4	Experimental Studies and Results		66
	4.4.1	Sensor Analysis	66
		4.4.1.1 Thinger.io Overview	67
	4.4.2	DHT11 Sensor Analysis	67
	4.4.3	MQ-2 Gas Sensor Analysis	70
	4.4.4	PIR Sensor Analysis	71
	4.4.5	Block Diagram of This System	75
	4.4.6	Sensor Recognition Process	76
	4.4.7	Real Life Implementation	81
4.5	Conclusion		84
	4.5.1	Future Work	85
	4.5.2	Discussion	85
References			85

4.1 INTRODUCTION

The importance of environment monitoring exists in many aspects. The condition environment is required to be monitored to maintain the healthy growth in crops and to ensure the safe working environment in industries, etc. Due to technological growth, the process of reading the environmental parameters became easier compared to the past days.

Internet of Things (IoT) is a rapidly increasing technology. IoT is the network of physical objects or things embedded with electronics, software, sensors, and network connectivity, which enables these objects to collect and exchange data. In this project, we are developing a system that will automatically monitor the industrial atmosphere and evaluate stored data and make decisions using the concept of IoT. IoT has given us a promising way to build powerful industrial systems and applications by using wireless devices, Android and sensors. The main contribution of this project is that it uses IoT in industries to monitor and control the Industry using various sensors and control units [1].

In recent years, a wide range of industrial IoT applications has been developed and deployed. In our project, we use different types of sensors that will help us to gather data and to inform about the current atmosphere where this project will implement. DHT11 is the Temperature and Humidity Sensor that features a temperature and humidity sensor complex with a calibrated digital signal output. By using the exclusive digital-signal-acquisition technique and temperature and humidity sensing technology, it ensures high reliability and excellent long-term stability.

The sensitive material of MQ-2 gas sensor is $SnO2$, which with lower conductivity in clean air. When the target combustible gas exists, the sensor's conductivity is higher along with the gas concentration rising. MQ-2 gas sensor has high sanctity to LPG, propane, and hydrogen, also could be used to Methane and other combustible steam, it is with low cost and suitable for different application [2].

PIR sensors are detecting sensors that use heat emitted to detect the place of an object or a living creature. Humans and warm-blooded animals emit heat and also infrared rays. This ray cannot be seen with human eyes because of low frequency.

Internet of Things (IoT)-Based Industrial Monitoring System

Infrared rays' frequency is between 3T and 430T; however, visible light frequency is between 430T and 750T [3]. IoT refers to the rapidly growing network of connected objects that are able to collect and exchange data using embedded sensors. It is nowadays finding profound use in each and every sector and plays a key role in the proposed environmental monitoring system too. IoT converging with cloud computing offers a novel technique for better management of data coming from different sensors, collected and transmitted by low-power, NODEMCUESP8266 microcontroller [4].

4.1.1 Background

The proposed project is implemented for monitoring the gas, temperature, motion, etc., and informs the connected people about the environmental condition through smartphone or portable device and gives an alert about the harmful conditions.

Bangladesh has taken an initiative to adopt the latest technological innovations, and some companies have already set up IoT laboratories in the country. Different countries have already adopted this technological innovation and 30 billion IoT devices are expected to be in the world by 2020, helping to improve performance [4].

IoT is a technology that connects machines to machines, giving city dwellers better performance and improving technological efficiency. This technological advancement will help find easy solutions for various problems, speciallywhile uploading sensitive information in the highly rated official websites.

In recent times, Bangladesh has nine sectors where IoT technology is used such as smart buildings, industrial automation, smart grids, water management, waste management, smart agriculture, telecare, intelligent transport systems, environment management, smart urban lighting, and smart parking.

IoT industry is the next wave of Internet technologies. IoT will fundamentally change how business and manufacturing will be done worldwide. It continues to spread across the home. IoT adoption reached some 43% of enterprises worldwide by the end of 2016. Total investment between 2015 and 2020 will be $6 trillion among both consumer and industrial IoT markets, with industrial IoT leading the growth [4].

4.1.2 Project Organization

In Section 2, we have discussed about the literature review that has to project relevant. In Section 3, we have discussed about the hardware and software we have used in this project. In Section 4, we have discussed about the experimental studies and results that we have conducted for our project purpose. In Section 5, we have discussed about the limitations and conclusion.

4.2 LITERATURE REVIEW

4.2.1 "IoT" and Its Smart Applications

This chapter discusses the concept of the project we are trying to implement, what kind of work has already been done on before, and how we are different at.

This chapter discusses about the concept of the project we have implemented, what kind of work has already been done before, and how we are different and more improved than those projects. In every organization, there is always an information desk that provides information, advertisement messages, and many notifications to their customers and staff. The problem is that it requires some staff that is dedicated to that purpose and that must have up to date information about the offer's advertisement and the organization. Due to IoT, we can see many smart devices around us. Many people hold the view that cities and the world itself will be overlaid with sensing and actuation, many embedded in "things" creating what is referred to as a smart world. Similar work has been already done by many people around the world. In the literature review, the IoT refers to intelligently connected devices and systems to gather data from embedded sensors and actuators and other physical objects. IoT is expected to spread rapidly in the coming years a new dimension of services that improve the quality of life of consumers and productivity of enterprises, unlocking an opportunity. Now this time, mobile networks already deliver connectivity to a broad range of devices, which can enable the development of new services and applications. This new wave of connectivity is going beyond tablets and laptops; to connected cars and buildings; smart meters and traffic control; with the prospect of intelligently connecting almost anything and anyone. This is what the GSMA refers to as the "Connected Life" [5].

4.2.2 Health Monitoring and Management Using Internet-of-Things (IoT) Sensing with Cloud-based Processing: Opportunities and Challenges

In this paper, the authors proposed the system health monitoring system and management using IoT sensing with cloud-based processing. They use a networking sensor to make possible the gathering of rich information indicative of our physical and mental condition. In this paper, they highlight the opportunities and challenges for IoT realizing this vision of future health care. Day by day technology will be improved so that it's our benefit. But here, the main context is IoT system proposed. This is the most important thing. Recent years have seen arising interest in wearable sensors and today several devices are commercially available for personal health care, fitness, and activity awareness. Technologically, the vision presented in the preceding paragraph has been feasible for a few years now. Yet, wearable sensors have, thus far, had little influence on the current clinical practice of medicine. In this paper, we focus particularly on the clinical arena and examine the opportunities afforded by available and upcoming technologies and the challenges that must be addressed to allow integration of these into the practice of medicine. Here we create the cloud and stored the data. This newly created data accusation sensing transmission, is processed, and finally cloud processing analysis visualization is done. But cloud let's processing and visualization are connected between the Internets. This model is the system architectures. It is impractical to ask physicians to pore over the voluminous data or analyses from IoT-based sensors. To be useful in clinical practice, the results from the analytics engine need to be presented to physicians in an intuitive format where they can readily comprehend the inter-relations between quantities and

Internet of Things (IoT)-Based Industrial Monitoring System

eventually start using the sensory data in their clinical practice. Finally, data gathered or inferred from IoT sensors span the complete spectrum of categories outlined in the previous paragraph and therefore an array of different visualization methodologies are required for effective use of the data. Eventually, this proposed system is effective for future technology effect [6].

4.2.3 IoT (Internet of Things)-Based Monitoring and Control System for Home Automation

In this paper, the authors explained about IoT-based monitoring and control system for home automation. A home automation system will control lighting, climate, entertainment systems, and appliances. It may also include home security such as access control and alarm systems. The home automation system uses the portable device as a user interface. This project was aimed for controlling home appliances via smartphones using Wi-Fi as a communication protocol and Raspberry Pi as a server system. The user here will move directly with the system through a web-based interface over the web, whereas home appliances such as lights, fans, and door locks are remotely controlled through the easy website [7].

4.2.4 Industrial Automation Using IoT

Automation is one of the increasing needs within industries as well as for domestic applications. Automation reduces the human efforts by replacing human efforts with a self-operated system. As we are making use of the Internet the system becomes secured and live data monitoring is also possible using the IoT system. IoT is achieved by using local networking standards and remotely controlling and monitoring industrial device parameters by using Raspberry Pi and embedded web server technology. Raspberry Pi module consists of the ARM11 processor and real-time operating system, whereas embedded web server technology is the combination of embedded device and Internet technology. Using embedded web server along with Raspberry Pi, it is possible to monitor and control industrial devices remotely by using a local Internet browser. They have developed new technologies that have allowed us to move from the first generation of the Internet into the current transition into the fourth generation. This generation has been propelled by the concept of the IoT [8].

4.3 HARDWARE AND SOFTWARE

In this project, we used different types of hardware and software. NodeMCU, ESP82663, Gas sensor, PIR sensor, and DHT11 were the main hardware. Arduino IDE, and Thinger.io Studio were mainly uses in this project.

4.3.1 Hardware

We used NODEMCU as the microprocessor for the system. To sense the LPG gas and to detect the motion uses PIR sensor and MQ2 sensor is used. All the Hardware components are used to setup the Generic breadboard.

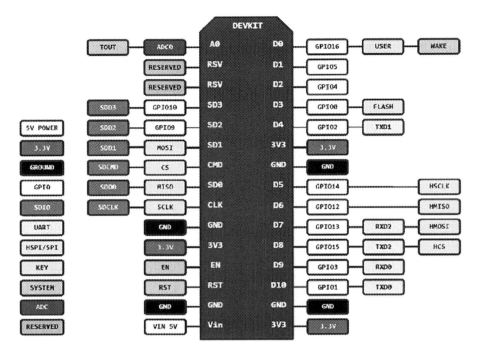

FIGURE 4.1 Node MCU pinpoint diagram [9].

4.3.1.1 NODEMCU

NodeMCU is an open-source IoT platform. It has a firmware that runs espresso system ESP8266 Wi-Fi SOC and ESP-12 module-based hardware. By default, the word "NodeMCU" refers to firmware instead of development toys. The firmware uses Lua scripting language (Figure 4.1).

The NodeMCU (Node Microcontroller Unit) is an open-source software and hardware development environment that is built around a very inexpensive System-on-a-Chip (SoC) called the ESP8266 [9].

4.3.1.2 ESP8266

The ESP8266 is a low-cost Wi-Fi microchip with full TCP/IP stack and microcontroller capability produced by Espresso Systems in Shanghai, China. It is a highly integrated chip designed to provide full Internet connectivity in a small package. ESP8266 is capable of functioning consistently in industrial environments, due to its wide operating temperature range. With highly integrated on-chip features and minimal external discrete component count, the chip offers reliability, compactness, and robustness (Figure 4.2).

ESP8266 contains a built-in 32-bit low-power CPU, ROM, and RAM. It is a complete and self-contained Wi-Fi network solution that can carry software applications as a stand-alone device or connected with a microcontroller (MCU). The module has built-in AT Command firmware to be used with any MCU via COM port [10].

Internet of Things (IoT)-Based Industrial Monitoring System

FIGURE 4.2 ESP 8266 module figure [10].

4.3.1.3 Gas Sensor

A gas indicator is a gadget that identifies the proximity of gases in a territory, regularly as a major aspect of a security framework. This sort of gear is utilized to distinguish a gas spill or different production and can interface with a control framework so a procedure can be naturally closed down. MQ2 for lowest cost monitoring solution with fast response (Figure 4.3).

MQ2 is one of the commonly used gas sensors in MQ sensor series. MQ2 gas sensor works on 5 V DC and draws around 800 mW. It can detect LPG, smoke, alcohol, propane, hydrogen, methane, and carbon monoxide concentrations anywhere from 200 to 10000 ppm.

4.3.1.4 DHT11

The DHT11 is a basic, ultra-low-cost digital temperature and humidity sensor. It uses a capacitive humidity sensor and a thermistor to measure the surrounding air, and spits out a digital signal on the data pin (no analog input pins needed). It's fairly simple to use, but requires careful timing to grab data (Figure 4.4).

FIGURE 4.3 Gas sensor [11].

FIGURE 4.4 DHT11 sensor.

They consist of a humidity sensing component, an NTC temperature sensor (or thermistor) and an IC on the back side of the sensor. For measuring humidity, they use the humidity sensing component which has two electrodes with moisture holding substrate between them.

4.3.1.5 PIR Sensor

PIR sensors allow us to sense motion, almost always used to detect whether a human has moved in or out of the sensors range. PIRs are basically made of a pyroelectric sensor which can be seen in Figure 3.5 as the round metal can with a rectangular crystal in the center, it can detect levels of infrared radiation (Figure 4.5).

FIGURE 4.5 PIR sensor [12].

Internet of Things (IoT)-Based Industrial Monitoring System

The output can be used to control the motion of door. Basically, motion detection use light sensors to detect either the presence of infrared light emitted from a warm object or absence of infrared light when an object interrupts a beam emitted by another part of the device [12].

4.3.1.6 Breadboard

A breadboard is a solderless device for temporary prototype with electronics and test circuit designs. Most electronic components in electronic circuits can be interconnected by inserting their leads or terminals into the holes and then making connections through wires where appropriate.

4.3.1.7 Power Board

We design and build a board to control the power of total device. This is the 10 V power supply. When the device is connected to the sensor for detecting data, that time power consumption is needed (Figure 4.6).

4.3.1.8 MCP3008

The Microchip Technology Inc. MCP3004/3008 Analogue to digital converter devices are successive approximation 10-bit analogue-to-digital (A/D) converters with on-board sample and hold circuitry. The MCP3004 is programmable to provide two pseudo-differential input pairs or four single-ended inputs. The MCP3008 10-bit Simple to-Advanced Converter (ADC) consolidates superior and low-power utilization in a little bundle, making it perfect for inserted control applications. The MCP3008 highlights a progressive guess register (SAR) design and an industry-standard SPI sequential interface, permitting 10-bit ADC ability to be added to any PIC microcontroller. The MCP3008 highlights 200 k examples/second, 8 information channels,

FIGURE 4.6 Power board.

FIGURE 4.7 MCP3008 [13].

FIGURE 4.8 Buzzer.

low-power utilization (5nA run of the mill backup, 425 µA normal dynamic), and is accessible in 16-stick PDIP and SOIC bundles (Figure 4.7).

MCP 3008 ADC converter device is used for data acquisition where multiple analog sensors are present and multiple sensors interface with this type of work.

4.3.1.9 Buzzer

A buzzer or beeper is an audio signaling device. Typical uses of buzzers and beepers include alarm devices, timers, and confirmation of user input such as a mouse click or keystroke (Figure 4.8).

The vibrating circle in an attractive signal is pulled in to the shaft by the attractive field. At the point when a wavering sign is traveled through the curl, it creates a fluctuating attractive field that vibrates the plate at a recurrence equivalent to that of the drive signal.

4.3.2 SOFTWARE

In this project we have used Arduino IDE to program NODEMCU. Arduino studio is used to connect to the web server. Here, a web server Thinger.io (https://thinger.io/) is used to connect ESP8266.

4.3.2.1 Web Server

A web server (thinger.io) has been used for this project. Because when sensor consumes the real time data that transmits the data through the web server. Thinger.io

Internet of Things (IoT)-Based Industrial Monitoring System

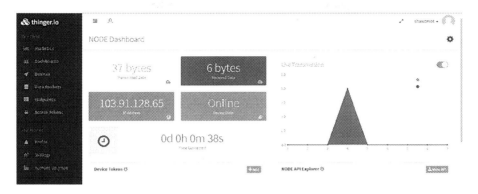

FIGURE 4.9 Web server.

platform is an open-source platform for the IoT; it provides a ready-to-use scalable cloud infrastructure for connecting things. There are some other important criteria that differentiate IoT platforms between each other, such as scalability, customizability, ease of use, code control, and integration with third-party software, deployment options, and the data security level [14] (Figure 4.9).

The Cloud Console is related to the management front-end designed to easily manage devices and visualize its information in the cloud.

4.3.2.2 Arduino Studio

The open-source Arduino Software (IDE) makes it easy to write code and upload it to the board. It runs on Windows, Mac OS X, and Linux (Figure 4.10).

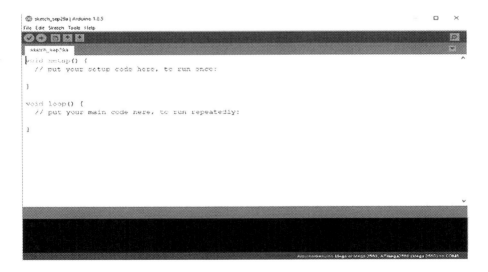

FIGURE 4.10 Arduino IDE.

4.4 EXPERIMENTAL STUDIES AND RESULTS

We did some sensor analysis before implementing this project. We also review some other projects that were related to this project and taking some idea and try to implement a new logic also cost-effective. We use different types of sensors for computing the environmental atmosphere and store data. The required data will store in a website using IoT and the website is Thinger.io server. (https://thinger.io/). By seeing those data and user can take the decision from this website. In our project, we used three sensors that we discussed in the previous section and here we represent those sensors data and interpret those data.

4.4.1 SENSOR ANALYSIS

Sensors play a key role in modern industrial plant operations. Nevertheless, the information they provide is still underused. Extracting information from the raw data generated by the sensors is a complicated task, and it is usually used to help the operator react to undesired events, other than preventing them. The complexity of modern industrial plants has called for equipment control automation that includes sensors for monitoring equipment behavior and remote-controlled valves to act upon undesired events. Plant automation physically protects plant integrity. However, it acts reacting to anomalous conditions. Equipment behavior set points within a working range are established and whenever the equipment behavior, such as equipment temperature, goes outside the designed range, an alarm is activated and a control equipment, such as a valve, is triggered to reset the equipment to the predefined working condition the complexity of modern industrial plants has called for equipment control automation that includes sensors for monitoring equipment behavior and remote-controlled valves to act upon undesired events. Plant automation physically protects plant integrity. However, it acts responding to abnormal conditions. Hardware conduct set focuses inside a working extent are built up and at whatever point the gear conduct for example hardware temperature goes outside the designed range an alert is initiated and a valve is activated to reset the hardware. The predefined working condition the unpredictability of present-day mechanical plants has called for hardware control computerization that incorporates sensors for checking hardware conduct and remote-controlled valves to follow up on undesired occasions. Plant robotization physically ensures plant uprightness.

Sensor analytics is the statistical analysis of data that is created by wired or wireless sensors. A primary goal of sensor analytics is to detect anomalies. The insight that is gained by examining deviations from an established point of reference can have many uses, including predicting and proactively preventing equipment failure in a manufacturing plant, alerting a controller in an electronic intensive care unit (eICU) when a patient's blood pressure drops, or allowing a data center administrator to make data-driven decisions about heating, ventilating, and air conditioning (HVAC).

Because sensors are often always on, it can be challenging to collect, store, and interpret the tremendous amount of data they create. A sensor analytics system can help by integrating event-monitoring, storage, and analytics software in a cohesive package that will provide a holistic view of sensor data. Such a system has three

Internet of Things (IoT)-Based Industrial Monitoring System

parts: the sensors that monitor events in real-time, a scalable data store, and an analytics engine. Instead of analyzing all data as these are being created, many engines perform time-series or event-driven analytics, using algorithms to sample data and sophisticated data modeling techniques to predict outcomes. These approaches may change, however, as advancements in big data analysis, object storage and processing technologies will make real-time analysis easier and less expensive to carry out.

4.4.1.1 Thinger.io Overview

In this section, we are going to depict the general architecture of the new open-source platform for deploying data fusion applications by integrating Big Data, Cloud, and IoT Technologies. Thinger.io is a new platform that is receiving interest by the scientific/technological community, finding projects in which this platform has been used successfully, including in the field of education. Thinger.io provides a ready-to-use cloud service for connecting devices to the Internet to perform any remote sensing or actuation over the Internet. It offers a free tier for connecting a limited number of devices, but it is also possible to install the software outside the cloud for a private management of the data and devices connected to the platform, without any limitation.

This platform is hardware agnostic, so it is possible to connect any device with Internet connectivity, from Arduino devices, Raspberry Pi, Sigfox devices, Lora solutions over gateways, or ARM devices, to mention a few. The platform provides some out-of-the-box features such as device registry; bi-directional communication in real-time, both for sensing or actuation; data and configuration storage, so it is possible to store time series data; identity and access management (IAM), to allow third-party entities to access the platform and device resources over REST/Web socket APIs; third-party Web hooks, so the devices can easily call other web services, send emails, SMS, push data to other clouds, etc. [12]. It also provides a web interface to manage all the resources and generate dashboards for remote monitoring. The general overview of this platform is available in Figure 4.1.

4.4.2 DHT11 Sensor Analysis

The DHT11 is a basic, ultra-low-cost digital temperature and humidity sensor. It uses a capacitive humidity sensor and a thermostat to measure the surrounding air and spits out a digital signal on the data pin (no analog input pins needed). It's fairly simple to use but requires careful timing to grab data. Following pin, the interface is given below and two libraries will be required to run this code. Download the zip file extracts the same and copies this in Arduino library folder. The pin diagram of DHT11 with NODEMCU is given below. It connects the 3.3 V with the NODEMCU. Some electrical characteristics of DHT11 are provided (Figure 4.11).

DHT11 Specification:

- Supply voltage: +5 V.
- Temperature range: 0–50 °C error of ±2 °C.
- Humidity: 20–90% RH ± 5% RH error.
- Interface: Digital. (Table 4.1 and Figure 4.12)

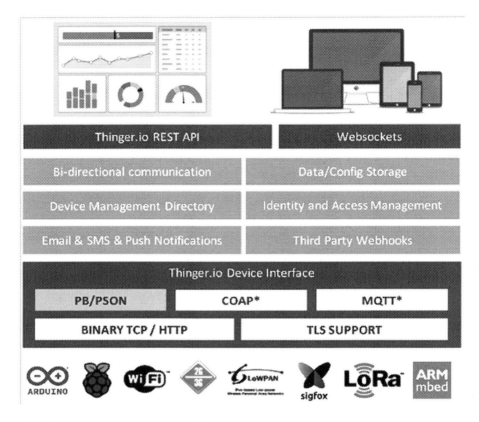

FIGURE 4.11 General overview of Thinger.io.

TABLE 4.1
DHT11 Specification Table [15]

Power Supply	Condition	Minimum	Typical	Maximum
Current supply	DC	3.3 V	5 V	5.5 V
	Measuring	0.5 mA		2.5 mA
	Average	0.2 mA		1 mA
	Standby	100 uA		150 uA
Sampling period	Second	1		2

This graph shows the value of temperature and humidity. The temperature and humidity change along with real time. DHT11 overall communication process along with real time is shown in Figure 4.13.

Communication process: Serial Interface (Single-Wire Two-Way) Single-bus data format is used for communication and synchronization between MCU and DHT11

Internet of Things (IoT)-Based Industrial Monitoring System

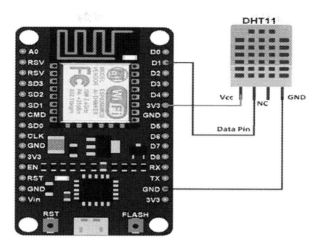

FIGURE 4.12 DHT11 connection diagram with NODEMCU.

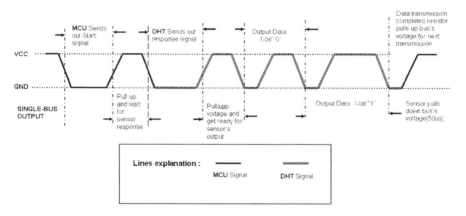

FIGURE 4.13 Overall communication process DHT11.

sensor. One communication process is about 4 ms. Once the start signal is completed, DHT11 sends a response signal of 40-bit data and trigger a signal acquisition.

Data format: 8-bit integral T data + 8bit decimal T data + 8-bit check sum. If the data transmission is right, the check-sum should be the last 8 bit of "8bit integral data + 8bit decimal data + 8bit integral T data + 8bit decimal T data".

When MCU sends a start signal, DHT11 changes from the low-power-consumption mode to the running mode. Once the start signal is completed, DHT11 sends a response signal of 40-bit data and trigger a signal acquisition. Users can choose to collect (read) some data. Without the start signal from MCU, DHT11 will not collect temperature information spontaneously. Once data are collected, DHT11 will change to the low-power-consumption mode until it receives a start signal from MCU again.

FIGURE 4.14 MQ-2.

4.4.3 MQ-2 Gas Sensor Analysis

MQ2 is one of the commonly used gas sensors in MQ sensor series. It is a metal oxide semiconductor (MOS) type gas sensor also known as **Chemi resistors** as the detection is based upon change of resistance of the sensing material when the gas comes in contact with the material. Using a simple voltage divider network, concentrations of gas can be detected. Ratio and concentration are nonlinear (Figure 4.14).

The MQ-2 gas sensor module detects gas leakage in home and industry. They are sensitive to a range of gasses and are used indoors at room temperature. The output of this sensor generates analog signal and can read through an analog pin of the Arduino. But, NODEMCU has only one analog pin, so we use MCP3008 chip for observing the digital output. This sensor can use in home or factory for sensing gas leak, suitable for gas, butane, propane, methane, alcohol, hydrogen, smoke, and other monitoring devices (Table 4.2).

This sensor has specific criteria for detecting the gas leakage criteria. An output criterion is high and low description. So some electrical specification is given below.

TABLE 4.2
Complete Specifications of Gas Sensor (Model MQ-2)

Operating Voltage	+5 V
Load Resistance	20 K(ohm)
Heat Resistance	33 ohm ± 5%
Heating Consumption	<800 mw
Sensing resistance	10–60 k ohm
Concentration Scope	200–10000 ppm
Preheat Time	Over 24 hour

Internet of Things (IoT)-Based Industrial Monitoring System

Specification of MQ-2:

- Size: 35 mm × 22 mm × 23 mm (length × width × height).
- Main chip: LM393, ZYMQ-2 gas sensors.
- Working voltage: DC 5 V.

Characteristics of MQ-2:

1. It gives a signal for output instruction.
2. Dual signal output (analog output, and high/low digital output).
3. 0 ~ 4.2 V analog output voltage, the higher the concentration the higher the voltage.
4. Better sensitivity for gas, natural gas.
5. A long service life and reliable stability.

NODEMCU connects the MQ-2 gas sensor. Gas sensor output shows the ADC converter (Figure 4.15)

This sensor gave real-time transmitted data through the web page which is connected to the thinger.io web page. When it detects the gas leakage that time output of this sensor's pick level became high otherwise pick level is low or remains in the steady state.

4.4.4 PIR Sensor Analysis

The construction and principle of operation of passive IR detectors (PIR detectors) of a large detection range. An important virtue of these detectors is highly efficient detection of slowly moving or crawling people. The PIR detector described here detects crawling people at a distance of 140 m. High signal-to-noise ratio was

FIGURE 4.15 Pin connection with NODMCU.

obtained by using a large number of pyroelectric sensors, i.e., by using a large number of detection zones (channels). The original electronic system for the PIR detector is presented in which DC signal amplifiers from pyroelectric signals are used. In order to ensure large detection ranges, a new method of signal analysis was used. The main elements of security systems are PIR detectors. In general, detectors operating inside buildings have a small detection range, small ranges of working temperature, and relatively simple algorithms of intruder detection. A passive infrared sensor (PIR sensor) is an electronic sensor that measures infrared (IR) light radiating from objects in its field of view. They are most often used in PIR-based motion detectors. PIR sensors are commonly used in security alarms and automatic lighting applications. Moreover, to detect slowly moving or crawling people, the lower limit frequency of a transfer band of PIR detector should be near zero. By fulfilling this condition, increase in low-frequency noise occurs causing a decrease in the next detector's sensitivity. A passive infrared sensor (PIR sensor) is an electronic sensor that measures infrared (IR) light radiating from objects in its field of view. They are most often used in PIR-based motion detectors (Figures 4.16 and 4.17).

➢ PIR Working Principle

The PIR sensor itself has two slots in it, each slot is made of a special material that is sensitive to IR. When a warm body like a human or animal passes through, it first intercepts one half of the PIR sensor, which causes a positive differential change between the two parts. Note that the PIR just uses a relatively simple sensor and definitely not a camera (Figure 4.18).

PIRs are called "passive" since they are not assisted by any "helpers" that for example would send some form or shape of "radiation" or "light" to help detect. It's purely based on what the sensor can pick from the environment, what is being emitted by objects.

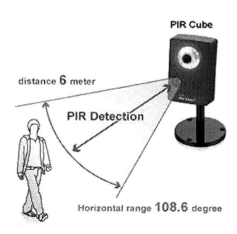

FIGURE 4.16 Object detection follows [15].

Internet of Things (IoT)-Based Industrial Monitoring System 73

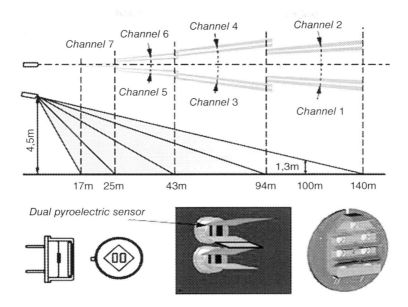

FIGURE 4.17 Detection zone of PIR deteector in horizontal and vertical planes [16].

FIGURE 4.18 Example of infrared radiation [16].

If the difference is too high then it will trigger – it detects "motion". This is done in a smart way, to avoid false positives caused for example by a brief flash or an increase in room temperature (Figure 4.19).

➢ The range of PIR sensor

PIR sensor detects a human being moving around within approximately 10 m from the sensor. It is an average value, as the actual detection range is between 5 m and 12 m. PIR are fundamentally made of pyroelectric sensor, which can detect levels of infrared radiation (Figure 4.20).

FIGURE 4.19 PIR sensor [17].

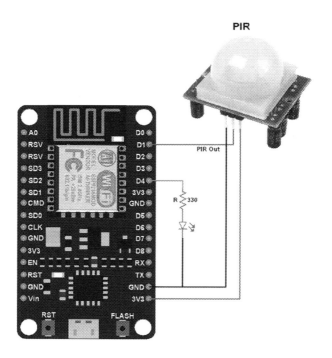

FIGURE 4.20 Pin connection with NODEMCU (PIR).

When sensors are connected with the NODEMCU, it observes the real-time data. When it detects any object that time PIR output will be high otherwise its output will be low or steady state (Figure 4.21).

To select Single Trigger mode, the jumper setting on PIR sensor set on LOW. In the case of Single Triggered mode, Output goes HIGH when motion is detected.

Internet of Things (IoT)-Based Industrial Monitoring System 75

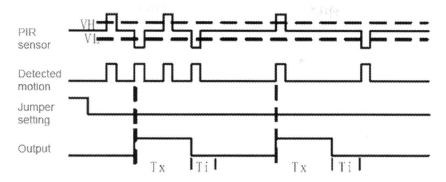

FIGURE 4.21 Overall communication system.

After specific delay, the output goes to LOW even if the object is in motion. The output is LOW for some time and again goes HIGH if object remains in motion. This delay is provided by user using the potentiometer. This potentiometer is on board of PIR sensor module. In this way, the PIR sensor gives HIGH/LOW pulses if object is in continuous motion. To select Repeat Trigger mode, the jumper setting on PIR sensor set on HIGH. In case of Repeat Triggered Mode, Output goes HIGH when motion is detected. The output of PIR sensor is HIGH until the object is in motion. When object stops motion, or disappears from the sensor area, the PIR continues its HIGH state up to some specified delay. We can provide this delay by adjusting the potentiometer. This potentiometer is on board of PIR sensor module. In this way, the PIR sensor gives HIGH pulse if object is in continuous motion.

➢ Changing Sensitivity and Delay time

There are two potentiometers on PIR motion sensors board: Sensitivity Adjust and Time delay adjust. It is possible to make PIR more sensitive or Non-Sensitive Enough. The maximum sensitivity can be achieved up to 6 meters. Time Delay Adjust potentiometer is used to adjust the time shown in the above timing diagrams. Clockwise Movement makes PIR more Sensitive (Figure 4.22).

4.4.5 BLOCK DIAGRAM OF THIS SYSTEM

We assemble the hardware layout by using sensor, Wi-Fi module, and others. The energy supply is mandatory for sensor reading. The whole power intake of the gadget is approximately 13–15 V. Now, connecting the sensor and look at the information Arduino studio serial display. After observing the serial display facts, all the sensors are linked using an Internet platform. When the Internet console signal is inexperienced which means the hardware is connecting the web platform. Finally, all sensors and other modules are linked and we observed the transmitted sensor statistics online.

Now, the block diagram of our project is given in the following Figure 4.23.

Thinger.io is able to take real-time data from all three sensors simultaneously and update it on the web page. The data are entered into the storage channel of the website and displayed graphically to the user. Following is a representation of one-time

FIGURE 4.22 Sensitivity changing switch [17].

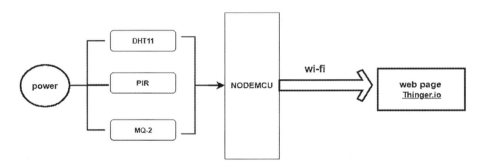

FIGURE 4.23 Project block diagram.

instance if the data are displayed to the user. The flow chart of this project is given below (Figure 4.24).

4.4.6 Sensor Recognition Process

Our project there are two parts. One is sensor reading and another is a web platform. All transmitted data are showing on the web platform. This is the most important thing in our project. Industry environment protection is an important sector for the security platform. We cannot monitor the whole working environment all the time manually. Using this platform, we can monitor the working environment and safety issues and can take proper steps regarding the worker's safety. This platform gives real-time data using IoT-based module and shows data through the web console or platform. So the user can observe data from anywhere where Internet is available. Our project is mainly based on IoT, so to monitor those types of workplaces we require an Internet connection to observe data. Here we show the web platform visualization (Figure 4.25).

Internet of Things (IoT)-Based Industrial Monitoring System

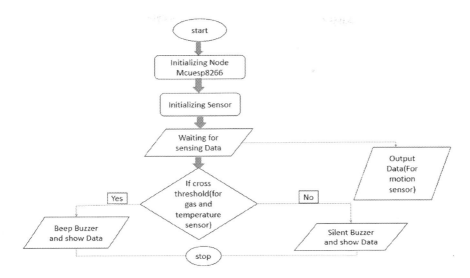

FIGURE 4.24 Flow chart.

FIGURE 4.25 Web dashboard.

We use the Thinger.io web platform. This is the free web console. Data we stored from the different sensors which were implemented in the required working area, those sensors are giving data about the working environment using the IoT module and those values were shown in the web platform. This web platform connects to the ESP8266 module and Thinger.io console just transmit the sensor values (Figure 4.26).

From 4.17 graphs, it can be seen that X-axis defines time and Y-axis defines ppm (gas limit), temperature, and binary output corresponding to the above figure from left to right. We observed that the output value MQ-2 gas sensor data gradually increases between 0 and 1000 ppm; and then decreases between 1000 and 0 ppm; and finally these data are saturated at 600 ppm. A normal clean environment data vary

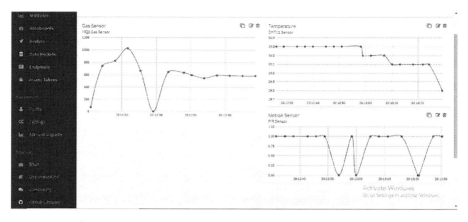

FIGURE 4.26 Sensor transmitting values in simple room.

from 150 to 600 ppm. In the temperature sensor when it crosses 30°C then it gives us an alarm. The output value of the temperature sensor DHT-11 did not vary so much and PRI sensor observe the whole working environment and gave an output based on the movement of the object or human as an oscillation from (Figure 4.27).

From 4.17 graphs, it can be seen that X-axis defines time and Y-axis defines ppm (gas limit). For a populated room, we observed that when the output value of MQ-2 gas sensors stays above 1000 ppm, then it can be considered a normal clean environment. Dangerous environment sensed as soon as drop below 1000 ppm is observed. It can also be detected when a sudden drop of gas limit is recorded. Such an instance triggers the alarm (Figure 4.28).

From 4.18 graphs, it can be seen that X-axis defines time and Y-axis defines temperature. Along with time, considering a fixed number of people in a room, the

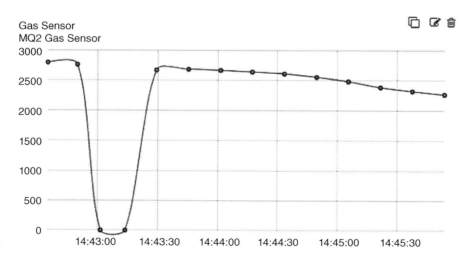

FIGURE 4.27 Gas sensor transmitting values in populated room.

Internet of Things (IoT)-Based Industrial Monitoring System

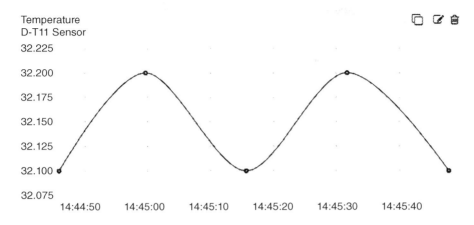

FIGURE 4.28 Temperature sensor transmitting values in populated room.

temperature remains fixed. As soon as the population increases with the entrance of new people in the room, the temperature increases, as seen at 14:45:00. When population decreases again, the temperature goes back to normal value again as seen at 14:45:15 (Figure 4.29).

From 4.19 graphs, it can be seen that X-axis defines time and Y-axis defines binary output. The PIR sensor records data with time, and when there is no motion in the room, the sensor reads a value of zero. When any motion is detected due any object in the room, the binary value rises to 1. PRI sensors observe the whole working environment and gave an output based on the movement of the object or human as an oscillation.

According to the atmosphere, we decided the three major levels for those sensors. MQ-2 gas sensor is a gas sensor for this sensor below 100 ppm is lower level and below 1500 ppm is middle level and above 1500 ppm is considered higher level and

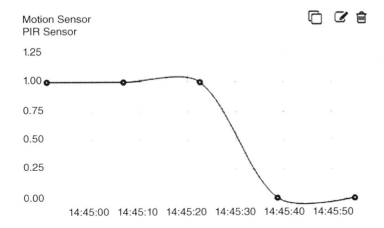

FIGURE 4.29 Motion sensor transmitting values in a populated room.

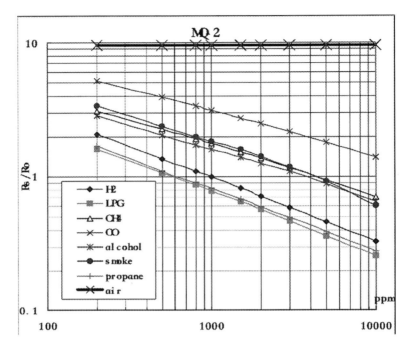

FIGURE 4.30 Sensitivity characteristics MQ-2 for different gases [18].

when working atmosphere gas level cross the middle level and entered into the higher level the buzzer will get on and give alert who monitor the working atmosphere [18]. Various companies have different working atmosphere. Chemical industry or pharmaceutical company mainly work in clean environment, and those type of environment are very gas sensitive (Figure 4.30).

Figure 4.30 represents the sensitivity characteristics of MQ-2 for different gases in different environment at 20°C temperature where humidity 65% and O_2 concentration is 21%. And normal resistance RL = 5 KΩ, R_0 = Sensor resistance at 1000 ppm, H_2 in the clean air. R_S = sensor resistance at various concentration of gases [18].

The Second sensor is DHT-11 which is temperature sensor and it also has three levels according to our working atmosphere. Some of factory work on low temperature and some of factory work on as usual normal temperature and most of the factory work on high temperature like chemical company and welding company or any factory where different types of product are making used by high-temperature machine that time we can set our three levels according to the working atmosphere. When middle level is crossed and entered into the higher level the system alert system, buzzer will get active and give signal to take proper action. And the last one PRI motion sensor which will do not give any alarm but this sensor all-time monitor the working atmosphere movement and show an oscillation in the output of the sensor.

Based on the interpretation of the diagrams, the operator decides his course of action. Signal lights present in the room will light up if harmful environment is

Internet of Things (IoT)-Based Industrial Monitoring System 81

detected. The workers and operator would then determine the conditions based on the color of the lights.

4.4.7 REAL LIFE IMPLEMENTATION

We implemented this project in our department because we did not get permission to implement this project in a large factory. So the result analysis based on data that we obtained from different implemented area (Figures 4.31 and 4.32).

We implement this project in switchgear lab and output of the results are represented in Figure 4.32 and Here, we can observe how data changed along with working atmosphere and how this project work and determine the level. He\she will decide his or her course of action based on the data and alarm. In this places. All the graphs in Figure 4.32, it represents the time in their X-axis and Y-axis for the first graph showing the gas sensor values has PPM count on its Y-axis. Second graph shows the temperature in Celsius in its Y-axis and third graph shows motion detection by PIR sensors in its Y-axis. The PIR sensor gives a binary output. It outputs 1 if some motion is detected and 0 if otherwise. The above Figure shows that MQ-2 gas sensor gave the lowest value 1075 ppm at temperature of 25.4 °C in the switchgear lab. The motion sensor observes the whole working environment and when a human being or any object are moving then the output of the PRI sensors show an oscillation based on the movement (Figure 4.33).

We also implement this project in Machine lab and output of this result represented in Figure 4.34. From 4.34 graphs shows X-axis define time and Y-axis defines ppm (gas limit), temperature, and binary output corresponding the above figure from left to right. We observe how data changed along with working atmosphere and how this project work and determine the level. Along with the time (X-axis), the output of the MQ-2 (Y-axis) is varying in ppm unit. Same as we saw a variation in the output

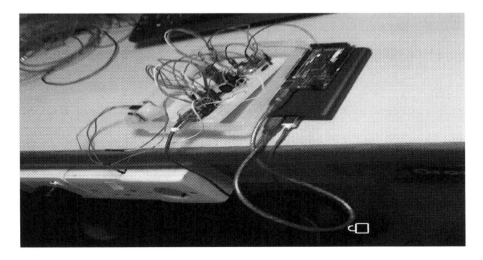

FIGURE 4.31 Real-life implementation in switchgear lab.

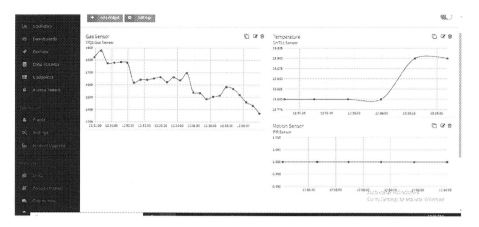

FIGURE 4.32 Real-life implemented Data in switchgear lab.

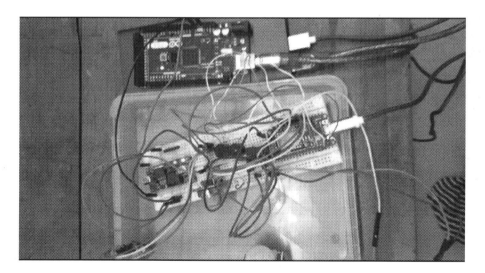

FIGURE 4.33 Real-life implementation in machine lab.

of the temperature sensor in the Y-axis along with time in the X-axis. Temperature went down along with time after some time output of the temperature sensor became stable. At last, the motion sensor observed the whole working environment and when a human being or any object were moving then the output of the PRI sensors shown an oscillation based on the movement. Then, he\she will decide his\her course of action based on the data and alarm. In this lab we did not get so much variation in the output, MQ-2 and DHT11 gave quite a similar value which we get in switchgear lab. The above figure shows that MQ-2 gas sensor gave the lowest value 1150 ppm at a temperature of 28.2 °C in the switchgear lab.

Internet of Things (IoT)-Based Industrial Monitoring System

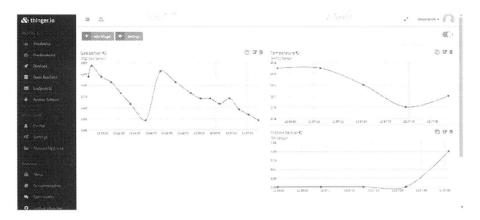

FIGURE 4.34 Real-life implemented data in machine lab.

We have used various gases to test out the gas sensor by some experiments. The first gas used was carbon dioxide (CO_2). The normal ppm of CO_2 is 300–750 ppm. In the presence of this gas, if the ppm range is exceeded, the buzzer sounds an alarm (Figure 4.35).

In Figure 4.35, it is observed that the sensor crosses the specified range at a time around 17:52:00. The abnormality remains for a certain duration, and throughout this duration, the buzzer will sound. Throughout the rest of the time, normal condition prevails and the ppm remains within a specified range.

The next gas used was LPG. The normal ppm of LPG is 500–900 ppm. In the presence of this gas, if the ppm range is exceeded, the buzzer sounds an alarm (Figure 4.36).

In Figure 4.36, it is observed that the sensor crosses the specified range at a time around 17:57:00. As long as the abnormality remains, the buzzer will sound. Throughout the rest of the time, normal condition prevails and the ppm remains within a specified range.

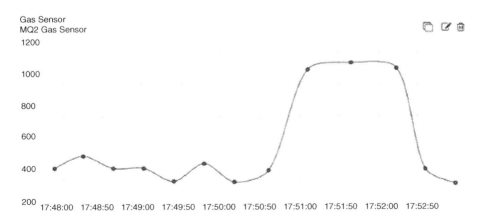

FIGURE 4.35 Implemented data for CO_2 gas.

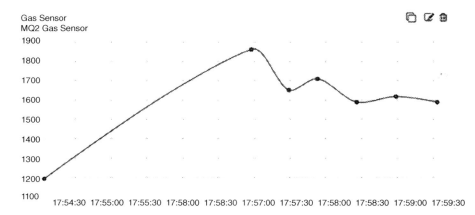

FIGURE 4.36 Implemented data for LPG gas.

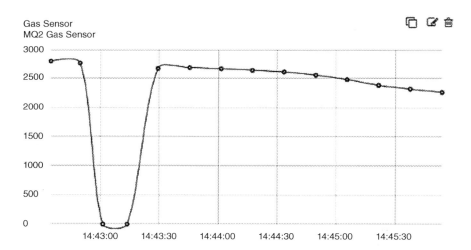

FIGURE 4.37 Implemented data for CH$_4$ gas.

The next gas used was methane. The normal ppm of *LPG* is 1500–2500 ppm. In the presence of this gas, if the ppm range is exceeded, the buzzer sounds an alarm (Figure 4.37).

In Figure 4.37, it is observed that the sensor reading goes below the specified range at a time around 17:43:00. As long as the abnormality remains, the buzzer will sound. Throughout the rest of the time, normal condition prevails and the ppm remains within a specified range.

4.5 CONCLUSION

The real-time analysis of the data provided by each of the sensors allows the user to have complete control over the environment of a system. The project being IoT-based allows remote monitoring of the target system even without physical presence.

4.5.1 Future Work

➤ Integrating more sensors for more specific data acquisition and analysis.
➤ Will be applicable in LPG gas services station in active situation.
➤ Will be used to provide Monitoring service to rural areas an affordable price.

Our project can be considered as a platform to develop in the field of IoT on the Monitoring system. In developing countries like ours, this kind of innovative and cost-effective project can improve the future of technology. So, we are looking forward to implement the project in order to make an impact in the new era of technology.

4.5.2 Discussion

The industrial monitoring system is designed specially to solve the cost-effective, accuracy and transparency problems in a highly secured approach. This system is more effective than the existing system since it uses an advanced controller for monitoring the environmental conditions and the controller collects the data from the sensor and those updated sensor values are written by the Arduino coding in particular text file. Using Arduino studio and the value is read and updated in the webpage. Based on the collected value, the respective action will be carried out. Our web page is created by Thinger.io. The whole of this system is user-friendly, because a user can see the environment information through the web page. Since this a prototype design, we would like to build a professional one in near future based on our project concept.

REFERENCES

[1] Al-Haija, Q. A., Al-Qadeeb, H., & Al-Lwaimi, A. (2013). Case Study: Monitoring of AIR quality in King Faisal University using a microcontroller and WSN. *Procedia Computer Science, 21*, 517–521.

[2] Chandramohan, J., Nagarajan, R., Satheeshkumar, K., Ajithkumar, N., Gopinath, P. A., & Ranjithkumar, S. (2017). Intelligent smart home automation and security system using Arduino and Wi-fi. *International Journal of Engineering And Computer Science (IJECS), 6*(3), 20694–20698.

[3] Cruz, T., Barrigas, J., Proença, J., Graziano, A., Panzieri, S., Lev, L., & Simões, P. (2015, May). *Improving network security monitoring for industrial control systems.* In: *2015 IFIP/IEEE International Symposium on Integrated Network Management (IM)* (pp. 878–881). Ottawa, ON, Canada, IEEE. doi:10.1109/INM.2015.7140399

[4] Deekshath, M. R., Dharanya, M. P., Kabadia, M. K. D., Dinakaran, M. G. D., & Shanthini, S. (2018). IoT based environmental monitoring system using arduino UNO and thingspeak. *International Journal of Science Technology & Engineering, 4*(9), 68–75.

[5] Deshpande, A., Pitale, P., & Sanap, S. (2016). Industrial automation using internet of things (IoT). *International Journal of Advanced Research in Computer Engineering & Technology (IJARCET), 5*(2), 266–269.

[6] Díaz, M., Martín, C., & Rubio, B. (2016). State-of-the-art, challenges, and open issues in the integration of Internet of things and cloud computing. *Journal of Network and Computer Applications, 67*, 99–117.

[7] Luis Bustamante, A., Patricio, M. A., & Molina, J. M. (2019). Thinger. io: an open source platform for deploying data fusion applications in IoT environments. *Sensors*, 19(5), 1044.

[8] Li, Z., & Gong, G. (2009, September). *DHT-based detection of node clone in wireless sensor networks*. In: *International Conference on Ad Hoc Networks* (pp. 240–255). Berlin, Heidelberg: Springer.

[9] Micko, E. S. (2007). *U.S. Patent No. 7,183,912*. Washington, DC: U.S. Patent and Trademark Office.

[10] Muller, K. A., & Mahler, H. (1991). *U.S. Patent No. 4,990,783*. Washington, DC: U.S. Patent and Trademark Office.

[11] Nayyar, A., Puri, V., & Le, D. N. (2016). A comprehensive review of semiconductor-type gas sensors for environmental monitoring. *Review of Computer Engineering Research*, 3(3), 55–64.

[12] Pavithra, D. & Balakrishnan, R. (2015, April). *IoT based monitoring and control system for home automation*. In: *2015 Global Conference on Communication Technologies (GCCT)* (pp. 169–173). Thuckalay, India, IEEE. doi:10.1109/GCCT.2015.7342646

[13] Prasad, S., Mahalakshmi, P., Sunder, A. J. C., & Swathi, R. (2014). Smart surveillance monitoring system using Raspberry Pi and PIR sensor. *International Journal of Computer Science and Information Technologies (IJCSIT)*, 5(6), 7107–7109.

[14] Ramya, V. & Palaniappan, B. (2012). Embedded system for Hazardous Gas detection and Alerting. *International Journal of Distributed and Parallel Systems (IJDPS)*, 3(3), 287–300.

[15] Shrouf, F., Ordieres, J., & Miragliotta, G. (2014, December). *Smart factories in Industry 4.0: a review of the concept and of energy management approached in production based on the Internet of things paradigm*. In: *2014 IEEE International Conference on Industrial Engineering and Engineering Management* (pp. 697–701). Selangor, Malaysia, IEEE. doi:10.1109/IEEM.2014.7058728

[16] Sharma, V., & Tiwari, R. (2016). A review paper on. "IOT" & It's smart application. *International Journal of Science, Engineering and Technology Research (IJSETR)*, 5(2), 472–776.

[17] Yavuz, S. O., Taşbaşi, A., Evirgen, A., & Akay, K. A. R. A. (1996). Motion detector with PIR sensor usage areas and advantages. *İstanbul Aydın Üniversitesi Dergisi*, 4(14), 7–16.

[18] Zhang, Z., Gao, X., Biswas, J., & Wu, J. K. (2007, July). *Moving targets detection and localization in passive infrared sensor networks*. In: *2007 10th International Conference on Information Fusion* (pp. 1–6). Quebec, QC, Canada, IEEE. doi:10.1109/ICIF.2007.4408178

5 Internet Working of Vehicles and Relevant Issues in IoT Environment

Rajeev Kumar Patial and Deepak Prashar

Lovely Professional University, Phagwara, India

CONTENTS

5.1 Introduction 88
5.2 VANETs Architecture 90
5.3 Communication Architecture 93
    5.3.1 In-vehicle Communication 93
    5.3.2 Vehicle to Vehicle Communication (V2V) 93
    5.3.3 Vehicle to Roadside Infrastructure (V2I) Communication 94
    5.3.4 Vehicle to Broad Cloud (V2B) Communication 94
5.4 VANETs Applications 94
    5.4.1 Traffic Signal 94
    5.4.2 Weather and Other Hard Conditions 95
    5.4.3 Vision Enhancement 95
    5.4.4 Assistance to the Driver 95
    5.4.5 Automatic Parking 95
    5.4.6 Information On Roadside Locations 96
    5.4.7 Entertainment 96
    5.4.8 Safety 96
5.5 DSRC Channels Access 97
5.6 A Case Study of Accident Detection on Road 98
5.7 Introduction to Clustering 100
    5.7.1 Data Aggregation 100
    5.7.2 Data Aggregation and Fusion Schemes in VANETs 100
5.8 Requirements of Security in VANETs 101
5.9 Menace to Availability 102
5.10 Menace to Authenticity 102
5.11 Issues in VANET 103
    5.11.1 Security 103
    5.11.2 Overheads Reduction 103

DOI: 10.1201/9781003102267-5

    5.11.3   Power Consumption 103
    5.11.4   Computational Time 104
5.12 Conclusion 104
References 104

5.1 INTRODUCTION

The *ad hoc* network of vehicles is meant for this purpose, and enables vehicles to have the capability to access the Internet by establishing communication between vehicles and the communication between vehicles and infrastructure. We need to switch over to the next technique involving wireless local area network (WLAN) systems (e.g., Wi-Fi). This system provides various advantages in terms of cost, performance, etc., as compared to other techniques. Also, recent research has evolved that mobile vehicles using Wi-Fi can access the Internet via Wi-Fi hotspots [1]. Therefore, Vehicular Ad hoc Network is defined as a network in which user vehicles are furnished with a service of information exchange with other adjacent vehicles wirelessly by making use of transceivers. For those vehicles that are not in a direct communication range, data exchange occurs via neighboring vehicles as shown in Figure 5.1 [2].

Vehicular ad hoc networks (VANETs) provide various interesting features because of attracting applications such as collision avoidance systems, driving assistance systems, etc. It allows people to access the Internet and to share information with other people by making use of active data streaming. It secures and supports the exchange of data which permits applications that can save lives such as position-based navigation applications, path combination based information applications, etc. Exclusively it also provides various advantages such as de-acceleration warning, congestion detection, public safety applications, traffic management applications, traffic coordination and assistance applications, traveler information support applications, broadband services, etc. (see Figure 5.2 and 5.3) [3].

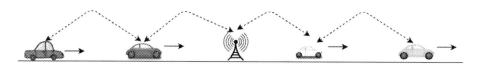

FIGURE 5.1   Segment of the road in VANETs.

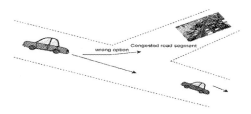

FIGURE 5.2   Congestion detection using VANETs.

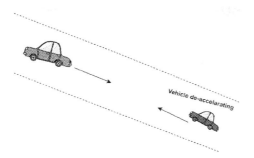

FIGURE 5.3 De-acceleration warning.

Irrespective of these numerous advantages VANETs, it is also associated with several challenging characteristics specifically in the field of privacy, security, and large-scale rapidly changing topology. Due to less-enhanced authentication features, there is a lack of authenticated data present in the network which increases the chance of malicious attacks and abuse of services; therefore, it can impose a threat to the public, passengers, and drivers. These challenges can be overcome by improving primary security essentials such as integrity, authenticity, and availability that need to be properly developed before implementing it practically [4]. Each vehicle in the network consists of a setup known as an onboard unit (OBU) used to integrate the functionality of vehicles in wireless communication, embedded systems, microsensors used for sensing various environmental conditions, and global positioning system (GPS) used to provide positioning information of the vehicles. Vehicles involved in this network can not only communicate with each other but also with other infrastructural units including roadside units (RSU) such as traffic lights, traffic signs, etc., as shown in Figure 5.4. This results in an improvement in the safety and driving experience of the participating drivers. The messages that are exchanged between vehicles are concerned to provide real-time traffic conditions to make the drivers aware of present knowledge of the driving environment to make a proper mechanism for rare situations as quickly as possible [5].

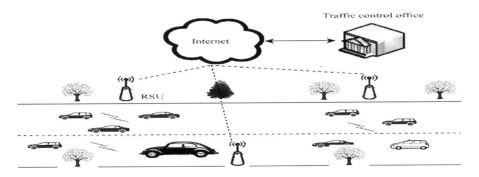

FIGURE 5.4 Overview of vehicular ad hoc network.

The main thing that is required is the integration of the VANET system with the IoT (Internet of Things) based environment as the deployment and the convergence of the protocols is very important for the effective utilization. Hence there is a need to understand the architecture and other issues of VANETs so that the convergence with the IoT systems can be done effectively.

5.2 VANETs ARCHITECTURE

This portion gives the detailed architecture of vehicular *ad-hoc* networks. Based on the domain the basic components of VANETs architecture are explained first followed by network architecture and then the communication architecture.

Referring to the IEEE1471-2000 [7, 8] and ISO/IEC42010 [9] architecture guidelines and standards, VANETs system architecture can be divided into the following three domains:

- Mobile domain.
- Infrastructure domain.
- Generic domain.

The mobile domain is further divided into two parts: the mobile device domain and the vehicle domain as shown in Figure 5.5. The mobile device domain consists of all types of portable devices smartphones and personal navigation devices. The vehicle domain consists of all types of vehicles like buses, cars, etc.

The infrastructural domain is further divided into two parts: the central infrastructure domain and the roadside infrastructure domain. The central infrastructure domain consists of vehicle management centers and infrastructure management centers, for example, traffic management centers (TMCs).

A generic domain is further divided into two parts: the Internet infrastructure domain and the private infrastructure domain [6].

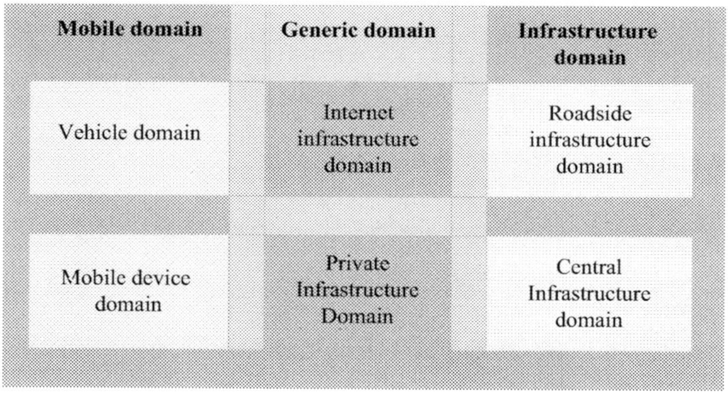

Mobile domain	Generic domain	Infrastructure domain
Vehicle domain	Internet infrastructure domain	Roadside infrastructure domain
Mobile device domain	Private Infrastructure Domain	Central Infrastructure domain

FIGURE 5.5 Domain architecture [6].

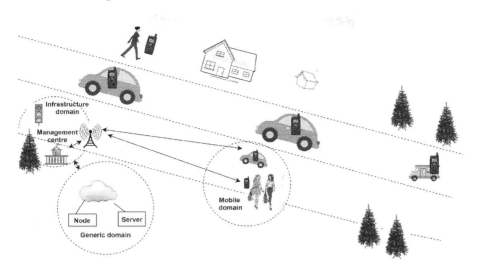

FIGURE 5.6 Domain architectural view of VANETs [6].

Though the growth of architecture of VANETs varies from one region to another. Figure 5.6 depicts the fundamental network architecture of VANETs. It has the following basic components:-

1. Onboard Unit (OBU) equipped on vehicles.
2. Road Side Unit (RSU) distributed over the infrastructure of the network.
3. Trusted Authority (TA)

Communication can take place between vehicles that are vehicle to vehicle and between vehicle and infrastructure. Each vehicle's onboard unit consists of a group of sensors to obtain information such as velocity, breaking information, etc. The roadside unit acts as a router to cover a wider area as compared to that of the area covered by vehicles. Vehicles are equipped with GPS to provide the information related to the positioning of the vehicles, electronic license plate (ELP) to provide the information related to the identification of the vehicles, radio detection, and ranging (RADAR) or light amplification by stimulated emission of radiation (LASER) can also be used to provide the information related to the positioning of the vehicles. The trusted authority types of equipment are equipped in the back end. The onboard units and roadside units communicate in a wireless manner using a Dedicated Short Range Communications protocol with an operating bandwidth of 75 MHz at 5.9 GHz frequency. Each roadside unit is connected which in turn is connected with the Trusted Authority (TA) using a wired connection. The Trusted Authority maintains the VANET system model.

1. On-Board Unit (OBU)

OBU is a transceiver equipped on vehicles for the exchange of information with the transceivers or OBU of other vehicles including the computational device and with the

RSU. The basic components of an OBU are resource command processor (RCP) for computation, storage, and retrieving of information, DSRC (Dedicated Short Range Communication) based on a radio technology IEEE 802.11p standard for wireless communication. OBU obtains its power from the battery of a car. Each vehicle is equipped with sensors such as GPS receiver, Event Data Recorder (EDR), Tamper Proof Device (TPD), forward, rear sensors to provide the input to the OBU, and speed sensors, etc. The sensors obtain the information from the surroundings of the vehicle, the GPS receiver provides information about the physical position of the vehicle. The EDR records the information of vehicle crashes or accidents. The TPD stores critical data including private key, identification proof of vehicles, and group key. The speed sensors obtain information related to the velocity of the vehicles. The forward and rear sensors collect the information related to the activities occurring around the vehicle by monitoring the front and backside of the vehicle. This collected information is then forwarded as a message to the neighboring vehicles by making use of the wireless medium.

2. Road Side Units (RSUs)

These are usually stationary; these devices are fixed on the sides of the road or the specific places such as road curves, parking places, etc. It consists of an antenna, sensors, processor, and transceiver. These units on the sides of the road provide the services to the vehicles such as road intersection which is used to control the traffic in that particular intersection and to reduce accidents. Each RSU makes use of a directional antenna or an omnidirectional antenna for wireless communication based on DSRC (dedicated short-range communication) IEEE 802.11p technology. To transmit a message to a particular location RSU makes use of a directional antenna. RSU possesses the capability of storing information obtained from OBU of vehicles and the TA (trusted authority).

3. Trusted Authority (TA)

It is meant for the registration of RSU, vehicle users, and the OBU of vehicles as shown in Figure 5.7. It verifies the authentication and authorization of OBU of

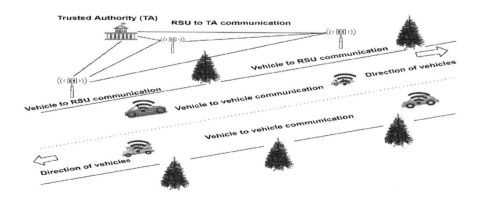

FIGURE 5.7 Network architecture of VANET system.

Internet Working of Vehicles and Relevant Issues in IoT Environment 93

vehicles and vehicle users to prevent the entry of malicious vehicles into the VANET system. It provides a high capability of storage and computation. It possesses the ability to uncover the real identity of OBUs when malicious messages are being broadcasted or when it shows a malicious behavior [10].

5.3 COMMUNICATION ARCHITECTURE

Communication taking place in VANETs can be classified into the following four categories. This type of architecture describes the functions of communications occurring in the VANETs system.

5.3.1 IN-VEHICLE COMMUNICATION

It refers to communication in the vehicle domain. It is responsible for collecting a piece of information related to the vehicle's performance. It detects the exertion, drowsiness, etc., of the driver meant for the safety of public, passengers, and the driver himself.

5.3.2 VEHICLE TO VEHICLE COMMUNICATION (V2V)

It refers to the communication between vehicles for the exchange of information including warning messages, etc., between them. This type of communication assists the driver and is shown in Figure 5.8.

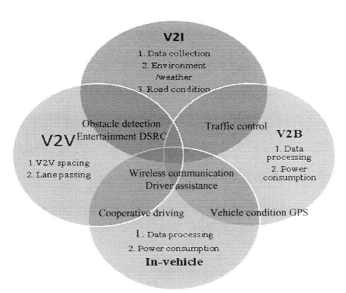

FIGURE 5.8 Key functions of each communication types.

5.3.3 VEHICLE TO ROADSIDE INFRASTRUCTURE (V2I) COMMUNICATION

It refers to the communication between vehicles and infrastructural units. It provides the information related to the current updates of weather, traffic, etc. It possesses the ability to monitor and sense environmental conditions.

5.3.4 VEHICLE TO BROAD CLOUD (V2B) COMMUNICATION

It refers to the communication between vehicles by making use of broadband (for example 3G or 4G or 5G). It includes traffic information, monitoring data, etc. This type of communication provides real-time and active assistance to the drivers and is also responsible for vehicle tracking [6].

5.4 VANETs APPLICATIONS

VANET-based communication can be used tremendously in numerous applications. It provides the capability to handle highly diverse requirements. Under broader sense applications of VANETs can be broadly classified into three categories which are safety directed, convenience directed and commercial directed. Safety directed applications monitor the surroundings of the road, vehicles, curves of the road, etc. Convenience application involves the management of the traffic. Commercial applications handle services provided to the driver. These services include the service of entertainment, web services, streaming of audio and videos, etc. Based on the representation and requirement of applications, certain applications are identified below:

5.4.1 TRAFFIC SIGNAL

It is possible to create communication from the traffic lights by making use of technologies embedded in VANETs. Slow or stop vehicle advisor (SVA) meant for safety applications provides a piece of information about the slow or motionless vehicle by broadcasting alert messages to their neighborhood as shown in Figure 5.9.

FIGURE 5.9 Traffic signal control.

To notify the road congestion, congested road notification (CRN) detects the road congestion, based on which journey and route are being planned. The toll collection at the toll booths without interrupting the vehicles is another type of application of VANETs. Vehicular networks are particularly useful in the management of traffic. However, VANETs for road tolling is widely deployed.

5.4.2 Weather and Other Hard Conditions

It consists of vehicle sensors such as wiper movement sensor, thermometer present outside to collect and update the weather information by making use of an application through DSRC (dedicated short-range communication). During an accident, when a vehicle is involved in an accident a warning message is being generated which would be broadcasted to the nearby traveling vehicles so that this information is passed on to the highway patrol for support. It also provides us the capability to notify the space availability in a parking lot for a specific geographical area by making use of Parking Availability Notification (PAN). It also possesses maps of the highway and urban areas to avoid traffic jams, conditions for an accident and to provide the shortest path in the critical situation leading to efficient usage of time.

5.4.3 Vision Enhancement

It provides a clear view of vehicles and obstacles and enhances the vision during the heavy fog conditions. It provides an ability to the drivers to recognize the existence of vehicles hidden behind obstacles, buildings and by other vehicles as shown in Figure 5.10.

5.4.4 Assistance to the Driver

VANETs provide the ability to support exercises in military driving by giving a piece of information to the drivers. Since vehicles may exhibit driving patterns in an abnormal way including dramatic change of direction, broadcast messages to inform cars that are present in their locality, therefore drivers can be warned by letting them know about the potential hazards to prevent accidents and get time to react. Thus, provides support to the driver in decision making.

5.4.5 Automatic Parking

This type of application shown in Figure 5.11 involves parking of a vehicle itself without any need for driver's interference. To perform such an action a vehicle

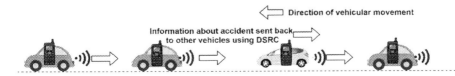

FIGURE 5.10 Obstacle detection.

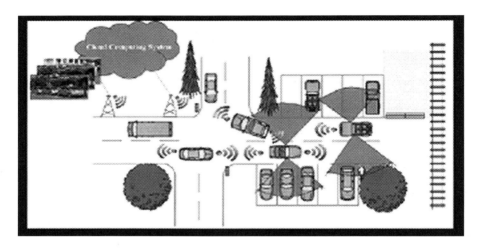

FIGURE 5.11 Automatic parking using VANETs.

needs an installation of a distance estimator, a sub-meter precision and a localization system.

5.4.6 Information On Roadside Locations

It provides the ability to search for roadside locations and also provides direction to the vehicles. Thus, passengers who are unknown to any particular location help them to find a specific location (for example, shopping centers, hotels, hospitals, etc.) in that area. It makes the use of GPS, database from the nearest roadside base station, sensors makes them able to perform such functions by calculating information.

5.4.7 Entertainment

Since several applications are involved to make the entertainment of passengers who are going to spend a long time traveling. Entertainment is provided in the form of Internet access, communication between passengers in the car's vicinity, games.

5.4.8 Safety

It involves collision warning, obstacle detection, and prevention, road condition warning, cooperative driving (such as lane merging warning), sign movement assistance, collision prevention of highway or railway, turn assistance, changing of lane warning shown in Figure 5.12(a) and (b).

Table 2.1 indicates the comparison among various applications based on latency, priority, message transmission range, and network traffic. Since the main applications of VANETs come under safety applications and non-safety applications. Safety applications involve the communication that is of broadcasting mode while non-safety applications are based on demand and request-response and include gaming, Internet surfing, multimedia, etc. [11, 12].

Internet Working of Vehicles and Relevant Issues in IoT Environment

(a)

(b)

FIGURE 5.12 (a) and (b) Traffic management control to provide safety.

TABLE 2.1
Comparison of Various VANET Applications [11]

Applications	Priority	Network Traffic	Allowable Latency (ms)	Message Range (m)
Life-Critical Safety	Class 1	Event	100	300
Safety Warning	Class 2	Periodic	100	50-300
Electronic Toll Collection	Class 3	Event	50	15
Roadside Service Finder	Class 4	Event	500	300
Automatic parking	Class 4	Event	500	300
Internet Access	Class 4	Event	500	300

5.5 DSRC CHANNELS ACCESS

It is very difficult to establish communication between the vehicles in the high-density city as well as highway scenarios due to the highly noisy environment around. Wireless communication around 2.45 GHz frequencies does not work efficiently in the VANET structure due to noise issues. Hence, it is important to increase the frequency of communication so that the effect of noise could be minimized. Therefore, dedicated short-range communication (DSRC), strive to achieve communication between vehicles at 5.9 GHz with reduced effect of noise on vehicular communication. 5.9 GHz. DSRC spectrum consists of seven channels of 10 MHz each, where one channel is reserved for control and six other channels are reserved for safety/service applications in VANETs. An additional 5 MHz is used as a guard band between channels. A structural review of DSRC at different frequency bands and different technologies are given below in Tables 2.2 and 2.3, respectively, to highlight the advantages of DSRC at 5.9 GHz.

Possible data rates for 10 MHz channels are 6, 9, 12, 18, 27 Mbps with a preamble of 3 Mbps as shown in the figure below

- Channel 172 is reserved for safety applications
- Channel 174 is reserved for applications that are shared by all

TABLE 2.2
Comparison of Wireless Technologies

	902–928 MHz Band	5850–5925 MHz Band
Spectrum	12 MHz	75 MHz
Maximum Range	300 ft	1000 m
Coverage	One communication area	Communication in overlapping zones
Data Rate	0.5 Mbps	6–27 Mbps
Interference Potential	High	Low
Minimum Separation	1500 ft	50ft
Channel Capacity	1–2 Channel	7 Channel
Downlink Power	Less than 40 dBm	Less than 33 dBm
Uplink Power	Less than 6 dBm	Less than 33 dBm

TABLE 2.3
Comparison of DSRC Technologies

	Satellite	Cellular	DSRC
Latency	10–60s	1.5–3.5 s	200 μs
Range	Thousands of kilometers	Kilometers	100–1000 m
Cost	Very expensive	Expensive	None

- Channel 175 is a combination of applications of channels 174 and 176
- Channel 176 is reserved for applications that are shared by all
- Channel 178 is the control channel, it supports all power levels, government control applications, broadcast messages
- Channel 180 is reserved for low power applications
- Channel 181 is a combination of applications of 180 and 182
- Channel 182 is reserved for low power applications
- Channel 184 is reserved for a high power service applications

Exact operational details of all infotainment applications are not yet standardized for most of the vehicular applications. Details of the few applications implemented using DSRC are given in Table 2.4. This information provides an initial understanding of VANET [8].

5.6 A CASE STUDY OF ACCIDENT DETECTION ON ROAD

Let's take an example of a road accident taken place on the highway, wherein there is a large density of vehicles as shown in the figure below. This road accident is detected by five nearby vehicles and each one of them propagated the message further on the highway. Subsequently, three other cars received these five messages and propagated all of them further on the highway. Later, these 15 packets are received by each of the

TABLE 2.4
Overview of Applications of VANETs [8]

Applications/Examples

Active Safety	Speed warning due to curvilinear road, warning due to low bridge, warning on traffic light violations, warning due to bad road condition, warning due to work zone, blind-spot warning, warning due to lane change, warning due to intersection collision, warning due to forward/rear, electronic brake lights in emergency, warning due to rail collision, warning due to pedestrians, pre-crash sensor warning, warning after crash, SOS service.
Public service	Warning due to approaching an emergency vehicle, reader for an electronic license plate, driver's license sensor, safety precautions for a vehicle and tracking of a stolen vehicle.
Driving experience	Assistant for merging of highways left turn and right turn, cooperative adaptive cruise control and glare reduction, crash notification warning or type of road surface to the center of traffic operation center, control of traffic with intelligence, more features in navigation and control, updated map availability and identification of parking spot.
Business/ Entertainment	Wireless diagnosis of information, flash in the software update, safety notifications, urgent repair notification messages, incorporating Internet service, entertainment notifications, management of a cluster of vehicles, availability of rental car, tracking of vehicles with hazardous material, collection of toll tax online, online payment of parking spots and gas.

red cars and disseminated the information to the car with green color and it received 45 packets. In this process as the message propagated from the source to the destination, Bandwidth consumption of the available channel also increased and when the message reached the red car, the channel was almost completely occupied by one type of message. This problem is called a broadcast storm problem in VANET and it can be overcome using clustering and data aggregation as depicted in Figure 5.13.

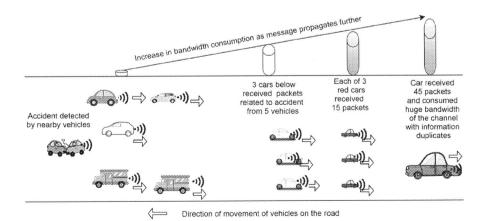

FIGURE 5.13 Data aggregation process.

5.7 INTRODUCTION TO CLUSTERING

Clustering is the process of structuring the unstructured data or it may be called the process of organizing vehicles into groups. The parameters taken to accomplish the task are average speed, standard deviation of speed, neighborhood degree of the vehicle, Tim to leave, trust factor, etc. It has been found that vehicles of similar speed and in the communication range of each other will form a cluster. These clusters can be of low-speed vehicles, intermediate-speed, or high-speed vehicles but always in communication range of each other. It is also noted the vehicles that are part of slow-moving cluster cannot form a group with a vehicle of intermediate speed or high-speed vehicle speeds, since connection of such vehicles cannot stay for a desirable duration to complete information transfer.

5.7.1 DATA AGGREGATION

It is the process of collecting data from multiple sources and presenting it in a summarized format. In the case of vehicular communication, multiple kinds of packets are generated by the vehicles as shown in table (overview of applications). Similar types of packets are aggregated together and propagated further on the road. Hence, instead of propagating multiple copies of the same packet, a single aggregated packet will be disseminated further on the highway, thereby saving enormous communication bandwidth.

The packet frame structure of VANET is given below [13].

First of all, the position of the vehicle is translated into $\{X-Y\}$ coordinates in integer meters. These coordinates are represented in sign-magnitude form as shown in Figure 5.14.

The speed of the vehicles is measured in km/s, with ΔS indicating the speed difference between vehicle speed and median of speeds of all vehicles of the cluster.

5.7.2 DATA AGGREGATION AND FUSION SCHEMES IN VANETs

The main goal of data aggregation algorithms is to gather and aggregate data in anefficient manner so that the validity of the information is ensured for a longer duration. In [14], authors define the in-network aggregation process as follows: In-network aggregation is the global process of gathering and routing information through a multihop network, processing data at intermediate nodes with the objective of reducing resource consumption (in particular energy), thereby increasing network lifetime.

Tree-based approach is the simplest way to aggregate data is to organize the nodes in a hierarchical manner and then select some nodes as the aggregation point or aggregators. The tree-based approach performs aggregation by

Frame type (1 Bit)	Time stamp (8 Byte)	Lattitude (8 Byte)	Longitude (8 Byte)	Data Section (Variable)	Signatures (28 Byte)	Certificate (56 Byte)

FIGURE 5.14 Aggregated frame.

constructing an aggregation tree, which could be a minimum spanning tree, rooted at sink and source nodes are considered as leaves.

The cluster-based approach is about the hierarchical organization of the network in tree-based approach. Another important technique to organize the network in a hierarchical manner is a cluster-based approach. In the cluster-based approach, the whole network is divided into several clusters. Each cluster has a CH which is selected among cluster members. CHs do the role of aggregator which aggregate data received from cluster members locally and then transmit the result to the sink. The advantages and disadvantages of the cluster-based approaches are very much similar to tree-based approaches.

5.8   REQUIREMENTS OF SECURITY IN VANETs

Requirements of security are the estimations that determine the extent of the secure network. The prevalent requirements of security include authentication, integrity, availability, confidentiality, non-repudiation, data verification, privacy. The absolute characterization of these security requirements is mentioned below as follows [5, 15, 16].

Authentication in VANETs ensures that the message is signed and sent by a registered authorized vehicle of a system without any modification. It guarantees that produced message is sent by an approved member of the system.

Integrity means the ability to not altering or modifying the sent information (message) without authorization by an unauthorized user from the time this information was sent. The modification of this information can be done either intentionally in a deliberate manner by an active attacker or unintentionally in an accidental manner, because of the faulty devices present in a vehicle.

Availability indicates that the system must be approachable to the verified member of the system anyhow devoid of attacks or being utilized in many ways by an attacker.

Confidentiality defines the capability of a system to stop and prevent the access of sent message or a piece of information to an unauthenticated and illegal users.

Non-Repudiation defines the ability in which a node whether in a sending mode or receiving mode can prove that the transmission or reception of information or message from a node and can prevent the node from denying its transaction [16, 17].

Data verification provides the capability to a node that message or information sent by the sending node is confirmed and the receiving node confirms this received information or message by a confirmation check to know whether this received information or message is ruined or not [5].

Privacy denotes that the data concerned with the driver i.e.; individual data must not be available to the unauthenticated vehicle [18, 19].

VANET attacks section summarizes the various types of attacks on the security of VANETs. Attacks in VANETs are widely divided into three main classes viz. attacks that create a menace to availability, attacks that impose menace to authenticity, those attacks that create a menace to confidentiality [3, 10, 20].

5.9 MENACE TO AVAILABILITY

Blackhole attack: This is the type of attack where an area is formed where nodes discard to take part in the network or where the participant node drops out from the established network resulting in the loss of data packets. Whenever the node drops out from the established network, all routes made by this node are broken which fails to propagate information or messages?

Denial of Service Attack: This is that type of attack where the attacker sends the number of dummy messages to other participating nodes and RSU to create congestion in the channels, which results in the reduction of performance and efficiency of the network. This type of attack is carried by network participants or network non-participants [21].

Distributed Denial-of-Service: This is similar to the denial-of-service attack. Here several malicious nodes take part in the attack on the confirmed node. Usually, these attackers participate per time slot in the attack.

Spamming: Here an attacker attacks the network by creating the presence of spam messages that demotes the latency of transmission and creating jam and delay of propagating information.

Malware: This type of attack involves the introduction of malware which includes viruses or worms in the network. This causes a critical reduction in the performance of VANET. This type of attack is usually executed by network participants and is introduced in the network when RSU and OBU software updating is performed.

5.10 MENACE TO AUTHENTICITY [22, 23]

Replay Attack: This is an active type of attack where the malicious attacker re-introduces the already received message or information in the network. Here the attacker saves the received packet and uses it further for replaying that is; sending the same packet, again and again, creating unnecessary route changes, stopping and traffic jams [14].

Masquerading: This is an active type of attack where an attacker personifies itself as a legitimate and confirmed participant of the network by providing fake IDs. It attacks the network. The attackers act as a "man in the middle" between two communicating vehicles and obtain the information, further can modify the information before forwarding it to the other vehicles [24].

GPS spoofing: In this attack an attacker creates the false reading of vehicles on GPS devices by making use of global positioning satellite simulator that generates stronger signals, which are more powerful signals than those generated by genuine satellite. This attack results in the traffic jam and fake position representation of vehicles [25].

Sybil attack: This type of an attack is an active attack, where a malicious attacker in the network pretends and claims of being multiple nodes and connects the network. Thus the operation of the network becomes insecure and the attacker possesses the ability to divide the network and can make the transmission of event-driven safety message restricted [26].

Internet Working of Vehicles and Relevant Issues in IoT Environment **103**

Tunneling: In this type of attack, the malicious attacker takes advantage of momentary loss of information about the position of the vehicle whenever a vehicle takes an entry in the tunnel. When this vehicle is about to receive the authentic information about the position the malicious attacker introduces the fake information into the OBU.

5.11 ISSUES IN VANET

This section discusses the leading objectives that effectively influence the quality of service in the operating network of VANETs. Among fundamental challenges, the following limitations are considered and are taken into consideration to overcome the degradation of the efficiency in the operating network of VANETs.

5.11.1 SECURITY

It implies that the individual data to be transmitted and received must be kept up secured against unauthorized vehicles and is required to be protected to prevent loss of confidentiality. Secure communication in VANETs is one of the noteworthy issues that is required to be resolved. To preclude the attackers and confirm the integrity and confidentiality of the exchanging information between vehicles, these vehicular nodes are required to be authenticated by following proper authentication procedures and the encrypted transmission of messages. Most appropriate mechanisms are required to be followed that specifies the standardized algorithm for the secure exchange of message from one vehicle to another without affecting the other reliable parameters of the network. In the IoT environment, the need of security of each device is a very crucial and important one for the overall security of the system.

5.11.2 OVERHEADS REDUCTION

It is one of the fundamental limitations that is needed to be omitted especially in the case of VANETS where the accuracy of time is essential to be maintained. Overheads not only creates an increased delay of communication but also makes the loss of energy and power. One of the most significant ways to reduce overheads is the introduction of clustering, possessing the potential to minimize the time of processing by decreasing the burden of the certificate authority. As the nodes in VANETs are a comparatively highly mobile and regular change of the topology of the network creates instability of nodes, thus providing a tremendous increase of overheads. For such maintenance of the stability of the network, clustering can reduce the overheads in a very significant way. The major challenge is the integration of the clustering approaches in the IoT-based systems once integrated with the VANET system.

5.11.3 POWER CONSUMPTION

One of the main factors that are required to be considered in VANETs, as the nodes are mobile, power usage must be limited to a lower value. One of the important parameters that affect power consumption is processing which in turn is related to the computational time. Optimizing the network to reduce the number of overheads, computational time will ultimately result in the reduction of power consumption. We need to perform such a mechanism that fulfills the significant requirements of security at a low usage of power so that there is significant progress over the standard alignments for the use of power without any noteworthy loss in the QoS (quality of service). Also in IoT systems low energy consumption should be the; prime concern for the long usage of the system pertaining to the harsh environment.

5.11.4 COMPUTATIONAL TIME

This parameter in VANETs is necessarily needed to be restrained as it affects the quality of service of the network. Computational time is required to be much condensed to provide accurate transmission and prevent accidents. Since computation time can be defined as the total time required for the accomplishment of a computational process. Several factors affect the computational time of VANETs. Numerous approaches are defined that can improve the performance by decreasing the computational time, in case of VANETs computational time is mostly affected by the decision of routing and the encryption and decryption involved in data transmission.

5.12 CONCLUSION

VANETs has widespread applications and also the emergence of many new applications based on IoT has also added advantage to it. There are many issues pertaining to the VANET deployment to its effective utilization but security is one of the prime concerns that needs to be look after. There are various attacks that are possible on the VANET systems and thus the requirement to mitiogate them is also one of the prime concerns. There are some other issues like power consumption, overhead, and computation time that need to be look after in the IoT environment as they are directly affecting the overall implementation and success of any VANET-based system once it is deployed in accordance to the IoT environment.

REFERENCES

[1] B. Zhang, X. Jia, K. Yang, and R. Xie, "Design of analytical model and algorithm for optimal roadside AP placement in VANETs," *IEEE Transactions on Vehicular Technology*, vol. 65, no. 9, pp. 7708–7718, 2016.

[2] C. Cooper, D. Franklin, M. Ros, F. Safaei, and M. Abolhasan, "A comparative survey of VANET clustering techniques," *IEEE Communications Surveys & Tutorials*, vol. 19, no. 1, pp. 657–681, 2017.

[3] L. Bariah, D. Shehada, E. Salahat, and C. Y. Yeun, "*Recent advances in VANET security: a survey*," In: *2015 IEEE 82nd Vehicular Technology Conference (VTC2015-Fall)*, Boston, MA, pp. 1–7, 2015.

[4] F. Qu, Z. Wu, F. Y. Wang, and W. Cho, "A security and privacy review of VANETs," *IEEE Transactions on Intelligent Transportation Systems*, vol. 16, no. 6, pp. 2985–2996, 2015.

[5] A. Luckshetty, S. Dontal, S. Tangade, and S. S. Manvi, "*A survey: comparative study of applications, attacks, security and privacy in VANETs*," In: *2016 International Conference on Communication and Signal Processing (ICCSP)*, Melmaruvathur, pp. 1594–1598, 2016.

[6] Review Article Vehicular Ad Hoc Networks: Architectures, Research Issues, Methodologies, Challenges, and Trends Wenshuang Liang, Zhuorong Li, Hongyang Zhang, Shenling Wang, and Rong fang Bie College of Information Science and Technology, Beijing Normal University, Beijing 100875, China, International Journal of Distributed Sensor Networks Volume 2015, Article ID 745303.

[7] Reference for DSRC: J. Guo and N. Balon. *Vehicular ad hoc networks and dedicated short-range communication*, University of Michigan, 2006.

[8] E. Schoch, F. Kargl, M. Weber, and T. Leinmuller, "Communication patterns in VANETs," *IEEE Communications Magazine*, vol. 46, no. 11, pp. 119–125, 2008.

[9] Rasmeet S. Bali, Neeraj Kumar, and Joel JPC Rodrigues, "Clustering in vehicular ad hoc networks: taxonomy, challenges and solutions," *Vehicular Communications*, vol. 1, no. 3, pp. 134–152, 2014.

[10] A Comprehensive Survey on Security Services in Vehicular Ad-Hoc Networks (VANETs) M. Azees 1, P. Vijayakumar 1,*, L. Jegatha Deborah 1 1 Department of Computer Science and Engineering, University College of Engineering Tindivanam, Melpakkam, Tamilnadu, India-604001.

[11] K. Ravi and K. Praveen, "*AODV routing in VANET for message authentication using ECDSA*," In: *2014 International Conference on Communication and Signal Processing*, Melmaruvathur, pp. 1389–1393, 2014.

[12] J. Dai, L. Pu, K. Xu, Z. Meng, Z. Liu, and L. Zhang, "*The implementation and performance evaluation of WAVE based secured vehicular communication system*," In: *2017 IEEE 85th Vehicular Technology Conference (VTC Spring)*, Sydney, NSW, pp. 1–5, 2017.

[13] K. Ibrahim and Michele C. Weigle, "*Optimizing CASCADE data aggregation for VANETs*," In: *2008 5th IEEE International Conference on Mobile Ad Hoc and Sensor Systems*, Atlanta, GA, USA, IEEE, pp. 724–729, 2008.

[14] M. Nema, S. Stalin, and R. Tiwari, "*RSA algorithm based encryption on secure intelligent traffic system for VANET using Wi-Fi IEEE 802.11p*," In: *2015 International Conference on Computer, Communication and Control (IC4)*, Indore, pp. 1–5, 2015.

[15] Y. Xie, L. Wu, Y. Zhang, and J. Shen, "Efficient and secure authentication scheme with conditional privacy-preserving for VANETs," *Chinese Journal of Electronics*, vol. 25, no. 5, pp. 950–956, 2016.

[16] S. F. Tzeng, S. J. Horng, T. Li, X. Wang, P. H. Huang, and M. K. Khan, "Enhancing security and privacy for identity-based batch verification scheme in VANETs," *IEEE Transactions on Vehicular Technology*, vol. 66, no. 4, pp. 3235–3248, 2017.

[17] S. K. Dhurandher, M. S. Obaidat, A. Jaiswal, A. Tiwari, and A. Tyagi, "Vehicular security through reputation and plausibility checks," *IEEE Systems Journal*, vol. 8, no. 2, pp. 384–394, 2014.

[18] Q. Li, A. Malip, K. M. Martin, S. L. Ng, and J. Zhang, "A reputation-based announcement scheme for VANETs," *IEEE Transactions on Vehicular Technology*, vol. 61, no. 9, pp. 4095–4108, 2012.

[19] M. Azees, P. Vijayakumar, and L. J. Deboarh, "EAAP: efficient anonymous authentication with conditional privacy-preserving scheme for vehicular ad hoc networks," *IEEE Transactions on Intelligent Transportation Systems*, vol. 18, no. 9, pp. 2467–2476, 2017.

[20] C. J. Wang, D. Y. Shi, and X. L. Xu, "*Pseudonym-based cryptography and its application in vehicular ad hoc networks*," In: *2014 Ninth International Conference on Broadband and Wireless Computing, Communication and Applications*, Guangdong, pp. 253–260, 2014.

[21] A. S. A. Hasan, M. S. Hossain and M. Atiquzzaman, "*Security threats in vehicular ad hoc networks*," In: *2016 International Conference on Advances in Computing, Communications and Informatics (ICACCI)*, Jaipur, pp. 404–411, 2016.

[22] H. Zhong, J. Wen, J. Cui, and S. Zhang, "Efficient conditional privacy-preserving and authentication scheme for secure service provision in VANET," *Tsinghua Science and Technology*, vol. 21, no. 6, pp. 620–629, 2016.

[23] K. Jeffane and K. Ibrahimi, *"Detection and identification of attacks in Vehicular Ad-Hoc NETwork,"* In: *2016 International Conference on Wireless Networks and Mobile Communications (WINCOM)*, Fez, Morocco, pp. 58–62, 2016.

[24] *Vehicular Ad hoc Networks (VANET) Engineering and simulation of mobile ad hoc routing protocols for VANET on highways and in cities Rainer Baumann*, ETH Zurich, 2004.

[25] B. Pradeep, M. M. M. Pai, M. Boussedjra, and J. Mouzna, *"Global public key algorithm for secure location service in VANET,"* In: *2009 9th International Conference on Intelligent Transport Systems Telecommunications, (ITST)*, Lille, pp. 653–657, 2009.

[26] J. T. Isaac, S. Zeadally, and J. S. Camara, "Security attacks and solutions for vehicular ad hoc networks," *IET Communications*, vol. 4, no. 7, pp. 894–903, 2010.

6 Adoption of Industry 4.0 in Lean Manufacturing

Nishant Jha and Deepak Prashar

Lovely Professional University, Phagwara, India

CONTENTS

6.1 Introduction 107
6.2 Related Work 109
6.3 Cyber-Physical Systems (CPS) 110
6.4 Importance of CPS Applications 112
6.5 Smart and Connected Products 113
6.6 Applications of Smart and Connected Products 115
6.7 Industry 4.0 115
6.8 Identification, Sensing and Communication 116
6.9 RFID and Networks 116
6.10 Middleware 117
6.11 Applications 118
6.12 Lean Manufacturing 118
6.13 Principles of Lean Manufacturing 118
6.14 Definition of Values 119
6.15 Mapping of Value System 119
6.16 Flow Establishment, Pull System and Origin of the Toyota
     Production System 120
6.17 Lean Learning Industries 122
6.18 Challenges of Lean Manufacturing 122
6.19 Conclusion 125
References 125

6.1 INTRODUCTION

Information and communication technology (ICT) has brought a progressive transformation to the globe over the past decades. There has been a movement toward more and more IT infrastructure, as well as the use of the Internet and product miniaturization [1]. Technological developments have resulted in new possibilities for smart objects to be linked through the Internet [1]. Industrial businesses must consider the immense impact of digitization and, in particular, of the Internet of Things (IoT). Smart products may offer major benefits, such as the establishment of a product-driven approach to production (i.e., the product takes the initiative during the execution of the plan); improvement of the entire product life cycle; improvement of the quality and efficiency of the product; and improvement of the next generation of the product [2, 3]. Industry 4.0 will become the data-intensive transformation of manufacturing (and related sectors) into an integrated collection of information, people, processes, facilities, systems, and also IoT-enabled manufacturing assets with the development and use of implementable data and information as a means and means of achieving sensible business and manufacturing innovation ecosystems and partnerships. Thus, Industry 4.0 is a holistic vision with distinct reference and structures, recognized mainly in so-called cyber-physical methods through bridging body output properties as well as electronic technologies. "We saw what happens when 3 billion people connect, next we'll see what happens when 20 billion computers connect" [4]. Such words by the vice-chair of GEs elegantly point to the immense potential of the fourth industrial revolution. Sensors have become inexpensive and efficient enough for the first time in the modern era; clouds are able to send, receive, and process vast volumes of data quickly enough and software is smart enough to draw concrete conclusions from data in real-time. Industry 4.0 is based on these technical advances. Industries should think about the emerging solutions as well as their opportunities and threats to uncover the promise of Industry 4.0. Based on the authors [5], it's crucial that systems that are different are totally integrated into one another to unlock the maximum possible value of Industry 4.0, or else, just 40–60% of the potential value could be captured. In addition, data collection needs to be strengthened, as the vast majority of data collected is not actually used in decision-making processes. The solution to that dilemma could be cloud-based computing systems such as GEs' Predix. Predix claims to have tremendous potential for streamlining processes, forecast future demand, minimize downtime of equipment, boost maintenance schedules, and increase production productivity while optimizing performance [6]. All of which will lead to more productive and effective activities, lowering costs and at the same time raising revenues. The possibilities seem infinite and the areas are relevant. From improved efficiency in jet engines, over real-time photos of a beating heart, to forecasting demand for electricity and changing the performance of wind farms accordingly. The Lean definition reflects the ongoing push toward greater efficiency. Lean is considered one of the most effective manufacturing principles worldwide and has found widespread acceptance in manufacturing firms. The essence of the idea is to minimize waste in production processes, thus improving productivity and reducing production costs [7–10]. Such a manufacturing paradigm would potentially benefit from the growth of Industry 4.0 because it uses streamlined processes and data analysis to identify and ultimately aim to eliminate waste in production. New technological advances are moving Industry 4.0 forward as with previous industrial revolutions. Cyber-physical systems (CPS) and the industrial Internet of Things (IIoT) are the most important of those emerging innovations for lean manufacturers. Fear of automation and other problems facing the manufacturing workforce makes it easy to overlook the possible positive cultural effect of these technologies. This new technology is an opportunity for lean manufacturers to reach the central goals of lean manufacturing: inspiring people to drive improvements. Technological developments have driven considerable changes in manufacturing productivity after the coming of the Industrial Revolutions [7–11]. The steam engine-driven industries in the nineteenth century, electrification and semi-automation in the

Adoption of Industry 4.0 in Lean Manufacturing

early 20th century, contributed to present-age manufacturing. Nevertheless, we're in the midst of the quarter era of technical advancement: the growth of innovative manufacturing technology referred to as Industry 4.0, a transformation run by nine fundamental technical advances. Under this technology, computers, sensors, and IT methods will likely be connected across the supply chain as a single unit. These connected systems (also referred to as cyber-physical devices) might interact with one another using standardized Internet-based protocols as well as analyze information to evaluate fault, configure themselves, and adapt modifications. Industry 4.0 enables information to be gathered as well as processed through computer systems, allowing quicker, much more adaptable plus more useful activities to generate products of better quality at decreased costs.

In turn, Industry 4.0 is going to increase market effectiveness, modification of marketplaces, promote quality product development, as well as eventually change the workforce profile shifting leading to the competitiveness between various nations and industries. Industry 4.0 is currently developing a new manufacturing area that will rely on the acquisition and exchange of data across the entire supply chain [1, 7–11]. At the same time, Industry 4.0 is a modern approach for connecting the physical world to the digital world. The word Industry 4.0 consists of CPS, IoT, Internet of Services (IoS), Robotics, Big Data, Cloud Manufacturing and Augmented Reality [11], and is the umbrella term for a modern industrial paradigm. We will address them in-depth in the coming parts with the difficulties and benefits of implementing Industry 4.0 with lean manufacturing.

6.2 RELATED WORK

According to Sundar et al. [12], "Lean" is the idea or philosophy of making a method more beneficial in different terms (mostly financial cases) by, for example, optimizing consumer value and reducing waste at the same time. The word "Lean" originates from "Lean meat", i.e., meat without any fat. It is a production methodology that is primarily customer-focused on process upgrades, where changes include optimization of time, human capital, money, efficiency, and quality, for example. Lean manufacturing sounds really good on paper, but real-life implementation is at least as critical as the hypothesis behind it. You also generate more value for the product as you minimize waste, which is, of course, satisfactory for the consumer, but there is still a possibility that the consumer will not be happy for different reasons such as design, requirements, etc. [13]. It is argued that the emphasis on the technological effects of new technologies within industry, rather than the broad effects of market and value sources, is not uncommon and has occurred in other disruptive efforts that were initially motivated by the technical effects they had [14]. This was evident in commonly embraced production initiatives such as Lean and Six Sigma, which, despite the lack of empirical evidence or awareness of their broader effect on business results, were implemented [15]. This is also supported by Dahlgaard-Park et al. [16] by acknowledging that much of the attention on Six Sigma is on the resources required to enforce the new program, with less attention on the market implications. With regard to Industry 4.0, Lasi, Fettke and Kemper [17] have argued that businesses must consider the organizational changes that will arise from the introduction of new business models and ways of operating by companies as a result of technological advances relevant to Industry 4.0.

Caputo, Marzi and Pellegrini [18] claim that, as opposed to the broader organizational ramifications, the literature on Industry 4.0 appears to concentrate mainly on the technical aspects. In the same vein, Roblek, Mesko and Krapez [19] note that the improvements needed by Industry 4.0 and the factors that triggered I4.0.0 have yet to be recognized by organizations. However, the author focuses solely on the technical improvements brought on by Industry 4.0, as opposed to any market implications. The increased range of product variants, shortened product life cycles, new production technologies and processes have resulted in a high level of competitiveness among manufacturing firms. There is a need for manufacturing companies to constantly develop and introduce changes in their production systems in order to remain competitive in order to meet these challenges [20]. In order to facilitate new products and processes, the development of production takes into account changes aimed at improving existing production systems or at creating completely new ones. The goal of production growth is for manufacturing companies to achieve efficient production processes [20]. Technical disciplines, such as business intelligence [21], data mining and knowledge discovery [22] and decision support systems [23], can be linked to data-driven decision-making. It is also important to consider factors relevant to decision-making and human judgment [24] in order to turn data into organizational meaning. Consequently, a framework is required where both technology and humans are taken into account.

A case study conducted by Ghobakhloo [25] discusses how the productive use of IT software as key innovations within Industry 4.0 will lead to changes in lean-digitized production systems. The case study conducted shows that a deep link with suppliers and customers is needed to enhance Lean Manufacturing practices. In order to promote electronic cooperation, this integration requires all supply chain participants to use IT-based technologies: it is claimed that effective Lean Manufacturing requires all parties to use integrated IT collectively to advance value chain activities and achieve wider integration. The case study shows that IT can enable manufacturing companies to generate market value through progressive lean-digitized manufacturing skills, and this can be seen as a good business strategy for survival in the era of Industry 4.0 as a hybrid manufacturing system. The use of Industry 4.0 has led businesses to reconsider digitalization and it is clear that Industry 4.0 is important for business strategies to be developed [26].

6.3 CYBER-PHYSICAL SYSTEMS (CPS)

So far, industry has undergone three revolutions: the first revolution began with the mechanization and development of machines that could use water and steam power; the second revolution occurred when human beings were able to produce and use energy and use it for mass production; and the third revolution began when electronics and automation became available in the industry. The fourth revolution is actually taking place through the invention of CPS [27]. The fourth revolution is made possible by three factors [27]. Computation: due to more efficient, more affordable and smaller processors, computation and digitization have become omnipresent. Sensing: sensing has become possible in almost every place and it is practically possible to measure. Connectivity: Virtually every computer will connect to the Internet or other networks in our world. The keyword for the fourth industrial revolution is the CPS. Computation and physical processes are integrations of cyber-physical structures. Embedded computers and networks track and regulate the physical processes, usually with feedback loops where computations are affected by physical processes and vice versa [28]. Data from sensors is used in traditional systems by the embedded controller to control the actuators and, as a result, the system. In CPS, there is a cyber controller parallel to the embedded controller that not only uses data from system sensors and the embedded controller, but also uses other data, such as historical data from the same system and data from similar devices and cloud management data, to optimally control the process and refine the process as time goes by. The cyber-physical framework would alter the human interaction which its physical environment [29] as the Internet altered the way people communicate with each other. Together with the IoT, Big Data, Cloud Computing, and Industrial Wireless Networks, the CPS is the central technology enabling the fourth industrial revolution, Industry 4.0. Smart manufacturing is becoming the focus of global manufacturing transformation along with developments in new-generation information technologies. In order to bridge the gap between design and development, realistic virtual models mirroring the real world are becoming necessary [30].

The cyber-physical device design is based on the architecture of 5C. These levels are structured to translate the big data obtained at the first level to the tiny yet precious data at the fifth level. The "communication level" devices are designed in the first stage to have the ability to self-connect and self-sense to collect data from three sources: sensors, controllers, and cloud data. Data from the connected devices are tracked and translated to the information in the "conversion stage" to identify the critical problems and monitor the health of the computer. This can be known as the system's self-awareness capacity; these data can be used in the next stage for future prediction of possible problems [27]. A cyber twin is created for each machine in the system in the "cyber level" and each machine can further examine its health by comparing its functionality with its cyber-twin, complex deep learning, and machine learning algorithms are used in this level to examine the health of the computer and its quality of operation. The outcomes of the previous levels will be provided to the user or decision-making software at the "cognition level" to decide on further actions according to the system situation, and finally at the "configuration level," the system can be reconfigured according to risk criteria [31]. The general overview of CPS can be divided into four main stages according to [32] as Figure 6.1.

Physical process: The regulation of physical processes and environment is a basic feature of CPS. It is often used to provide input on any past steps taken by the CPS, and to encourage proper future development. Networking: This stage is concerned with the aggregation and distribution of data. Computing: The purpose of this stage is to reason and analyze the data obtained during the monitoring process in order to verify if the physical process meets the preset parameters. Actuation: This step performs the behavior discovered during the computation process [33].

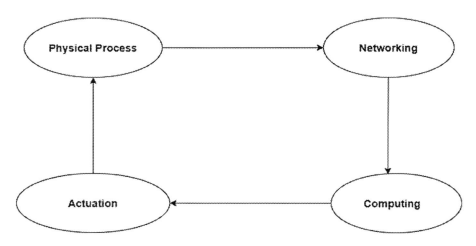

FIGURE 6.1 Four-step theory of CPS. (Adapted from [32].)

6.4 IMPORTANCE OF CPS APPLICATIONS

According to [33], the implementation of CPS in manufacturing systems gives rise to a smart factory, which is an important concept under Industry 4.0. Vertical integration, as the environment for vertical integration in the Industry 4.0 sense, means the implementation of a highly versatile and reconfigurable smart factory [33]. The plant is responsible for the actual production of raw materials and semi-finished products. During production and management, numerous physical or information subsystems are involved at the boundary of a factory [33]. Traditional organizations are also organized into a functional unit hierarchy. As a result, problems arising at the interface boundaries are often given less priority than the units' short-term outcomes [33]. Smart factory adoption will greatly boost this scenario. Since the smart factory leverages the network of interconnected system knowledge to operate in a highly effective, agile and scalable manner, the overall architecture can be split into three key sections: modules, machines and production systems [33]. Different levels of comprehension and openness of the factory can offer us each of these items [33]. The advantages of smart factories over traditional ones are as follows [34].

- Transparency: Big data offers real-time, total, and effective data on all facets of a smart factory, according to [35]. According to [36], since invisible problems can occur due to machine degradation, part wear, etc., in comparison to visible problems, although operators and factory managers are not aware of them, factory-wide transparency is one of the most important goals of the future factory. In the factory, the implementation of CPS will bring clarity to the factory, allowing us to measure performance metrics relating to equipment, goods and systems [37].
- Staff friendly: The smart factory should provide the human workers with personalized modifications so that the machine adapts to the human work cycle. In addition to this, maintenance and diagnosis are made simpler

Adoption of Industry 4.0 in Lean Manufacturing 113

with the help of big data analytics, efficient software tools, and more comfortable and versatile interface measures [33]. In industrial automation, mobile devices such as smartphones and tablets have already made inroads.

- Customer integration: Via intelligent compilation of the ideal production system, which factors account for product properties, prices, logistics, protection, reliability, time and sustainability considerations, the smart factory can conduct optimized individual customer product production [33]. It allows the goods to be tailored to the unique and individual needs of the customer [33].
- Energy efficient: In addition to the high prices, the use of large quantities of raw materials and energy by the manufacturing sector often presents quantities of raw materials, and energy often presents a range of environmental and security risks. Smart factories will produce improvements in resource productivity and quality to solve this issue [33].

While the importance of the production maintenance process is beyond question, in the past, it was commonly branded as a "necessary evil" by owners of organizations. Maintenance is accountable for the management of overhead expenses, including the price of manpower, supplies, instruments, etc. [33]. In addition, it has been found that divisions in refineries, maintenance and operations are always the largest and each occupies around one-third of the overall staffing. In addition to these estimates, most of the indirect impacts of maintenance on efficiency have contributed to the confirmed belief that maintenance has a lower return rate than any other major budget item [33]. Thus, despite the undeniable contribution of maintenance to overall efficiency, most manufacturers have found it to be an unavoidable cost center [33]. Different generations of maintenance and total quality management are shown by Figure 6.2.

As a result, several techniques have been built to promote the implementation of maintenance management in the industry over the years [33]. This is shown in Table 6.1.

6.5 SMART AND CONNECTED PRODUCTS

Three core elements are divided into a smart and connected product: physical components, "smart" components, and networking components. The mechanical and electrical parts of a product (e.g., the engine block, tires, and automobile batteries) are physical components. Sensors, microprocessors, data storage, sensors, applications, an integrated operating system and an improved user interface (e.g., engine control unit, anti-lock braking system, automatic wiper rain-sensing windshields and touchscreen displays) are 'intelligent' components. Software also allows a single physical system to function at a number of levels. In addition, it is also normal for hardware components to be replaced entirely by software [1]. The ports, antennas, and protocols that enable wired or wireless communication with the product are connectivity components. In other terms, these are of three kinds, one to one, one to many, and many to many [1].

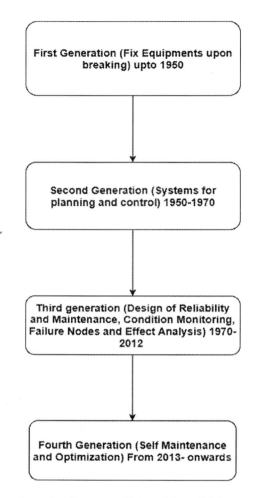

FIGURE 6.2 Generations of maintenance. (Adapted from [38].)

TABLE 6.1
Total Quality Management Contributions and Objectives (Adapted from [33])

Objectives	Contributions
Volume of Production	Unplanned breakdowns minimized, Enhanced availability of equipment, Improved actual time of production
Quality	Reduced problems from dysfunctional manufacturing, Reduced defects due to the frequently slow phase of maintenance
Cost	Reduced life-cycle costs and reduced waste in the course of maintenance
Safety	Improved environment in the workplace and no accidents
Delivery	Improving delivery performance by improving reliability and promoting Just-In-Time efforts with reliable equipment

Adoption of Industry 4.0 in Lean Manufacturing 115

6.6 APPLICATIONS OF SMART AND CONNECTED PRODUCTS

An entirely new set of functions and capabilities is possible through communication and knowledge built into a product. The four types of applications or capabilities listed below are applicable [1]

- It is about the monitoring of the product's state, function, and external environment. Sensors and external sources of data can handle this. The advantage is that if there are changes in circumstances or results, the product will send a warning. In addition, operating characteristics and the history of a product may be monitored. This knowledge helps to explain how the object is used. Through multiple items and long distances, tracking can be achieved.
- Software that is built into the system or exists in the product cloud performs the control. The product may react to changes in its condition or environment through remote commands or algorithms. The advantage is that product output is tailored into a new level. This allows product interaction to be managed and personalized in various new ways.
- In conjunction with the ability to manage the process of the product, monitoring data enables product efficiency to be improved in various ways. Through applying algorithms and analytics to real operational data or historical data, this can be done. The gain is an increase in the product's performance, consumption, and quality. It is also reported that real-time product condition monitoring data and product management capability helps companies to maximize service by conducting preventive maintenance when failure is imminent and remotely conducting repairs, thereby reducing production downtime and the need to dispatch repair staff.
- By integrating monitoring, access and control, autonomy can be achieved with capabilities for optimization. Products can adapt to their environmental conditions, recognize the need for their own service and the preferences of the customer. Benefits include less operator control needs, enhancing safety and promoting service in remote areas. Autonomous products also allow coordination with other products and systems to be possible.

6.7 INDUSTRY 4.0

The concept of Industry 4.0 is that the world is now at the brink of a fourth industrial revolution. The phenomenon is based on the ability to allow a new kind of smart end-to-end development and facilitation with current technologies, such as machine-to-machine communication, sensor technology and big data analysis [39]. The concept is that these technologies would encourage a new form of intelligent and self-optimizing machinery that synchronizes with every aspect of the value chain, from raw material development, design or ordering to servicing and recycling [39]. Industry 4.0 is affecting the growth of information communication technology (also known as ICT) in the industry. The vision is to link the real world to a Digital Twin,

116   Industrial Internet of Things

to collect and evaluate data in order to better predict results and to help decisions, automations, and the entire value chain [39]. Industry 4.0 is a highly complex and radically distributed network environment from a machine perspective, composed of a large number of nodes, generating and consuming information through a cloud-based system [39]. Due to the heterogeneity of the topic and its interdisciplinary nature, defining the theoretical delineation of IIoT is a challenge. Three paradigms can be used in realization of IIoT i.e., identification, sensing and communication, Middleware, and Applications [39]. These are explained in upcoming sections.

6.8   IDENTIFICATION, SENSING AND COMMUNICATION

The first element of the chosen categorization is aimed at allowing technology for data collection and information communication. These are the wireless "things" which drive forward the vision of the industrial revolution. The information that is transmitted is at the level of the device, both from "sensing" the world and communicating this externally, but also from digesting data sent by other devices. In this context, wireless technologies are allowing data transfer and playing a key role as deployment is growing. The increase in the deployment of transmittable sensors indicates that a decrease in scale, weight, energy consumption and cost will lead to a new way of incorporating "radios" into almost all objects, which is truly an integral part of the IIoT concept [39].

They must be active parts of the entire enterprise, the information processes, and connect and communicate with each other by ingesting and exchanging data from their environment in order to investigate how these "stuff" can function in Industry 4.0, thus reacting autonomously to the information sensed and exchanged at some stage. This may be by running procedures that, with or without direct human involvement, trigger acts and build greater value [39]. Radio frequency identification sensors (RFID) are the most debated kind of "smart sensors". These have existed for a while and are mostly used (highly embraced by metro systems and modern keys) as identification tags [39].

6.9   RFID AND NETWORKS

These identification tags have been developed to sense other RFID tags as the technology has advanced, and perform actions based on personalized triggers. For example, the presence of tags in the surrounding area and providing a reading of the current environment can be a request for this. It can then be used in real-time, without the need to be in sight, to track objects, enabling an accurate virtual representation of the real world, creating a Digital Twin. The application scenarios are numerous with this kind of virtual map, ranging from logistics to tracking complex and complicated bottlenecks [39]. Reviewed from a technological point of view, the RFID tag is a small chip with an antenna that can relay information using induction. This implies that the power supply is the reader and not the sender, and the RFID tags run without a battery. Since these are very simple in terms of how to sense outside environments, other than nearby RFID tags and potential readers, the low cost is why this is an important piece of technology and in this paradigm can still be considered a

Adoption of Industry 4.0 in Lean Manufacturing

breakthrough [39]. For Industry 4.0, the sensor network is important for the partnership of RFID systems and other sensor styles. These can be configured to integrate temperature and sensor information that the RFID device does not support in order to close the distance between the real world and the virtual world. The combination of the different types of sensors would result in a sensor network consisting of large numbers of intelligent nods that collect, process, and analyze useful data gathered from vast environments [39].

6.10 MIDDLEWARE

Middleware is a software component that enables long-distance communication and data transfers of large volumes, offering more complicated processing and storage of data. Experts believe that the centralization of computing resources is feasible. It is the design and software where all the sensors should be centralized and installed on, becoming the middle stage between data collection (from sensors and machines) to applications (programs, analysis, presentation, automation systems, etc.) becomes centralized and installed on. One main aspect of the middleware is its function of shielding the specifics of various technologies from the diversity of the entire IoT system to allow the programmer to focus on specific applications. Therefore, not presenting all data gathered in an unsorted manner, but more specified and sorted data storage to allow the creation of a particular application tailored to the desired type of outcome [39]. In recent years, middleware has gained prominence in the handling of big data and complex infrastructures, as the new technology has increased its processing and computational power and thus the possibilities of creating a sophisticated middleware system [40]. The ability to uniquely recognize devices, functionalities and environments when offering a standard range of services is one of the key problems with regard to the middleware part [40].

A multi-layered structure is proposed by many researchers to be able to integrate a large number of different types of data sources, as it can assist in the standardization of data processing and storage, known as Service Oriented Architecture (SOA) [39]. The SOA's approach and growth in popularity is largely due to its concepts of decomposing more well-defined data and components to complex and monolithic structures. This implies for the whole business environment, with uniform interfaces and protocols. The theory is based on the idea that all business data should be correlated with procedures, workflows, and object action [39]. This should promote communication and cooperation between the various sections, resulting in less time required in the fluctuating environment to adjust itself. Software and hardware reuse is another popular aspect of the SOA approach. Since the architecture does not need a particular technology or software for the implementation of the application, both the technological components and the hardware [39] may be reused. The unified infrastructure to support both storage and analytics in real-time is another significant feature of middleware. Cloud-based storage and computing technologies are a recent and growing trend. The newly achieved ability to manage unprecedented volumes of data, but also the ability to display the data on multiple platforms, not limited by venue, is one of the key advancements in cloud computing [39]. Smart applications must be built on the middleware for all this to work.

6.11 APPLICATIONS

The last phase of the Industry 4.0 implementation is the framework. On top of the architecture, this is the visual component of the framework, exporting to the end user the entire virtual representation, data, and analysis. The potential visualizations and implementations differ as this model is highly oriented around the user, as it is not as limited by recent technical progress, but depends on the desired visualization of the application. To separate from the middleware, this paradigm is essential, as it is used for all functionalities, taking advantage of the data structure and streams to allow top layer applications. It can be defined as the perfect integration between distributed systems and applications [39]. To be able to have creative apps, the best approach seems to be a single system across platforms. The possibility to exchange knowledge through the application framework should also be an important feature of the application system. The philosophy of cloud data storage and computing should embrace this centralized form of framework, since it is one of its main capabilities. Nevertheless, businesses need to consider the emerging developments and the threats and opportunities that emerge in order to unlock the potential of Industry 4.0. In addition, data collection needs to be enhanced because the vast majority of the data obtained is not actually taken into account in decision-making processes [39].

6.12 LEAN MANUFACTURING

Lean manufacturing is a method of production that offers a way to specify value, line up value-generating actions in the best sequence, execute these operations without interruption if anyone requests them, and execute them more and more efficiently. Lean manufacturing is a concept established in Japan, where the main managers of Toyota concentrated on waste reduction in production systems; Taiichi Ohno described the first seven types of waste in an assembly line, which will be discussed ahead. They started incorporating lean thinking in all production-related activities in order to reduce this waste and turn it into value development, trying to generate more value with less resources and moving closer to supplying consumers with what they really need [41]. Lean manufacturing is a management model that focuses on minimizing the losses of the manufacturing processes while optimizing the end client's value development. The strategy of cost reduction should not be mistaken; applying lean output means converting waste into value for the consumer, not just reducing waste [41].

6.13 PRINCIPLES OF LEAN MANUFACTURING

There are five principles of lean manufacturing [42]. These are as follows: (1) Identify value according to the product family from the viewpoint of the end user. (2) For each product family, define all the steps in the value stream. (3) Make a continuous flow of value. (4) Let clients pull value from the next upstream operation. (5) Pursue perfection: Pursue a relentless development method that aims for perfection (Figure 6.3).

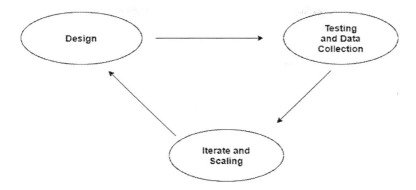

FIGURE 6.3 Principle of lean manufacturing. (Adapted from [42].)

6.14 DEFINITION OF VALUES

Defining the importance during lean implementation is also one of the most difficult tasks. Most manufacturers want to sell what they already make, and they typically fall back on the same ideas when they try to redefine value [41]. The same applies to users, who just know how to ask for versions of what they already have. Another explanation why it is difficult for businesses to define value is that the flow of value typically goes through different locations. The value to fit their own needs is specified by each region, but adding them is not the value of the entire output. This may be an instance of how individual activities can be optimized, but resources are wasted by the full production system [41]. Identifying the value requires manufacturers to connect with consumers, not only outside the organization, but with the various upstream and downstream operations. Acceptance of the task of redefinition is crucial because it is also the key to gaining more clients, and this is very necessary. As one of the lean concepts is to guarantee the company's properties, such as the workers, if they minimize the waste and create more from less, the company will have to find more buyers and profits faster. The most critical step in the definition of value, after the commodity has been identified, is the determination of the target cost on the basis of the resources available, which is the key to waste reduction [41].

6.15 MAPPING OF VALUE SYSTEM

Identifying the value development of the production process is the first step toward lean execution. It is crucial to identify which processes generate value for the consumer, which ones add non-value but are important, and which ones add non-value. The value stream mapping (VSM) is one method used to define these value flows. The VSM allows businesses to identify the activities that generate value and highlights where changes can be made in a given workflow or process [41]. The approach is based on one basic premise: the tasks must be observable and well defined in order to be able to enhance and reduce waste (they must have a specific beginning and end) [41]. The waste in the production system has to be, in addition to the non-value operations, has to be discovered. Waste is something other than the minimum quantity of equipment, materials, components and working time that is completely necessary for the product or service to add value [43].

6.16 FLOW ESTABLISHMENT, PULL SYSTEM AND ORIGIN OF THE TOYOTA PRODUCTION SYSTEM

When you look at most of the manufacturing processes, the commodity spends most of the time waiting to be processed; it is only in a few moments that value is actually obtained by the product [43]. The sum of waiting time is a waste as previously seen, it is important to calculate a flow along the processes in which the product spends the minimum waiting time and the maximum acquiring value to prevent this waste. To create flow, there are three techniques. Once value is established, the first is to retain the concept of the product and remember what its particular design and order is. The second goal, which makes the first one possible, is to forget the conventional boundaries of tasks and employment so that all the barriers to the continuous flow of a particular product or product family can be removed. The third one is to reduce the waiting period, back lows and scrape along the production [41], reconsider the job and activities. One of the pillars on which Lean Theory is based is the Pull system. In this method, the customers are the ones asking for the object. Instead of manufacturing the machines at the highest cost, they manufacture to meet the downstream customer's demand. The downstream customers are therefore the ones who have begun the production order for their predecessor station [43]. With this system, the amount of WIP present in the system can be decreased, achieving the ideal of a single-piece flow. In order to enforce these systems, it is important to use Kanban cards, for example. The Kanban may be cards or signals of various kinds that indicate when to start working for the predecessor station, but in any case it is a way to pull the machine to produce when necessary [41].

Workplace optimization is the traditional second step in applying the lean approach; this optimization is referred to as housekeeping (5S) [43]. These represent sorting, set in order, shining, standardizing and sustaining. Various tools for Lean Manufacturing is summarized in Table 6.2.

The Toyota Way consists of Continuous Development and Consideration for People, two elements. Continuous progress consists of Challenge, Continuous Improvements (Kaizen), and Respect and Cooperation consists of Go and See and Respect for People [44]. Challenge suggests that it should always challenge the existing status in order to push changes. Questions such as: Why do we get the outcomes we receive? Why is the new method of development looking the way it is? Why are the elements of the job in this series being done? In order to constantly question the status quo in order to push improvements, they should be questioned [44].

Continuous Improvement (CI) implies that one should strive to improve at all times. "Always a better way" inside Toyota is a commonly used slogan. The Japanese term Kaizen is used when talking about CI. Kaizen means successful (zen) shift (kai) and is used interchangeably with CI. It is stressed when teaching CI to concentrate on the many minor changes rather than the few big improvements. The explanation is that the many small changes are easy to find and to come up

Adoption of Industry 4.0 in Lean Manufacturing

TABLE 6.2
Tools for Lean Manufacturing (Adapted from [1, 45])

Defects	Poke Yoke: It is a way of removing the possibilities of manufacturing faulty goods and involves two distinct types of structures. First, the warning systems which, if there is a deviation from a norm, send a signal. Second, control systems that interrupt a computer if there is a deviation from a norm. Standard Work: Tasks should be outlined in detail to decrease the likelihood of one task being overlooked. Also, it will be simpler for various operators to perform the job.
Overproduction	SMED: If changeovers are quick, it can be cost-effective to produce smaller batches and thus reduce the time a batch is waiting to be allocated to an order. Two rules exist: 1) When the system is running, perform as many tasks as possible. 2) Reduce the time required to do off-line tasks on the computer. Kanban: It is a self-managed production mechanism in which the end of a process (customer order) shows how much commodity material is consumed and therefore needs to be replenished. The signal can be obtained at any point on the value stream (e.g., refill for a box with screws).
Waiting	Takt-time: The time the consumer requests the product (e.g., the consumer needs 730 products a year, the tact has to be at least two products a day) Line-balancing: The aim is to align a production or assembly line in a manner that has equivalent cycle times for all sequential workstations.
Unused Talent	Training: Lean tools and practical problem solving should be educated by everyone in an organization.
Transportation	Production lines: Transforming the layout of the plant into production lines where a product flows in the same order through the same workstations. Work cells: In a U-shape, narrow production lines are designed to minimize transportation, motion and maybe operators as well. And 5S
Inventory	One-Piece-Flow: Restrict batches to one product. Outcomes of a reduced amount of work in the process. Work-cells and Kanban
Motion	Standard work and 5S
Excessive processing	Process map: Six symbol types (process phase, delay, inventory, decision, calculation, and transport) flowchart. Observing to get details in the process about hidden waste.
Muri (Overburden in machine environment)	Preventive and Autonomous Maintenance
Muri (Overburden in Human Environment)	Jidoka Principles, Standard Work and 5S
Mura	Heijunka, Modular Designs, Standard Work, SMED, 5S and more.

with, with a low implementation risk, they are also typically very simple and relatively inexpensive. However, major changes can take a longer period to be introduced, during which period there are enhancements. A major change is often potentially more complex, more costly and thus implies a higher risk. Where a significant change does not work as planned or does not give the expected results [44]. Another idea is Go and See that is where the value-adding process is being conducted, management should be present. This implies that managers do not hide in their offices in manufacturing, but rather are out on the shop floor to thoroughly try to understand the current circumstances and to try to provide help in terms of

122   Industrial Internet of Things

problem-solving and questioning the current best standard of driving CI [44]. Effective implementation of Lean needs creating a community that embodies the ideals of the Toyota Way. The right Lean thought and culture will not flourish without them, and instead there will be an emphasis on unique Lean resources that will be difficult to maintain in time [44].

6.17 LEAN LEARNING INDUSTRIES

As a learning environment, learning factories create a reality-based development workplace. The trainees will experiment, test and explore the technologies applied here, as well as get used to the workstation's distribution and efficiency. The main aim of these training environments is to improve the skills and abilities of the learners effectively in challenging or unfamiliar circumstances. In addition to focusing on production areas, these training courses generate value by developing the skills of factory workers along the value chain at all hierarchical levels in various technical and organizational areas of operation. Therefore, the context of the learning factories can be said to be largely didactic, but with a clear real application [41]. The Lean model follows these five fundamental principles suggested by [39] since its origin: (1) Specify value; (2) Recognize value stream; (3) Prevent value flow interruptions; (4) Let customers pull value; (5) Once again start chasing excellence. Lean's meaning, however, is constantly evolving and over time the five basic principles have been changed in [39], for instance, take a practical view of Lean as a set of 22 management tasks divided into four sets, just-in-time (JIT), Total Quality Management (TQM), Total Preventive Maintenance (TPM) and HRM. With a view to achieving a waste-free production cycle [39], JIT tries to constantly minimize waste. According to [39], the inventory of work-in-process (WIP) and excessive delays in inflow time are two major sources of waste and can be minimized by collecting output flow-related methods such as "lot size reduction, cycle time reduction, rapid exchange of WIP inventory reduction techniques and cellular format implementation, output process re-engineering, and bottleneck reversals." However, for these tools to allow JIT, certain pre-conditions have to be set. TQM focuses on the following nine core concepts, including cross-functional material design, process management, quality control of manufacturers, customer participation, knowledge and feedback, dedicated leadership, strategic planning, cross-functional training and employee involvement, with the aim of continually improving and maintaining the development of quality products [39]. A strong commitment to TQM [39] was one reason Japan's automotive industry outperformed its Western counterparts after World War II. The concept of lean manufacturing is shown in Figure 6.4

6.18 CHALLENGES OF LEAN MANUFACTURING

- Customer satisfaction: Lean manufacturing sounds really good on paper, but the reality in real life is at least as important as the philosophy behind it. You also generate more value for the product as you minimize waste, which is, of course, satisfactory for the consumer, but there is still a chance that the consumer will not be happy for different reasons such as design,

Adoption of Industry 4.0 in Lean Manufacturing

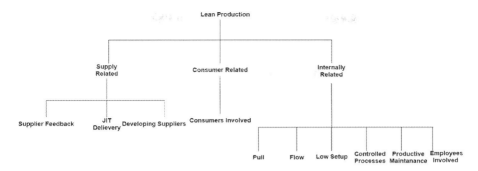

FIGURE 6.4 Concept of lean manufacturing. (Adapted from [39].)

requirements, etc. Lean manufacturing processes are very dependent on the performance of suppliers, which makes the process vulnerable to supply chain disruptions. Thus, if the consumer is not happy, due to delays in supply chain deliveries [13], this may lead to long-lasting marketing issues for the company.

- Productivity: Lean manufacturing theories contribute to a significant increase in productivity, but the downside is that the cost of productivity rises. The fourth stage in the 5S is about standardizing the method, but it will cost a lot of money in the implementation phase to make a work process structured as the theories explain. Another problem with a structured work process is that the implementation of new features or technology into the process can be challenging because you would have to take into account all the applicable requirements that have already been implemented [13].
- Quality: Implementing the lean theories of manufacturing can lead to better product quality, but this comes with high implementation costs in the same way as efficiency. If a corporation does not follow the Lean guidelines, a substantial portion of the physical plant structures would have to be dismantled. You may also need to invest in new and more effective equipment for your process and train staff to implement the new philosophy [13].
- Delivery times: The Just-In-Time concept is a philosophy that can be applied if you don't want to have a surplus stock that just sits down and focuses more on keeping consumer orders as valuable as possible. If workers want to strike, if there are transport delays, or if there are any quality errors that can interrupt production, which can lead to an economic crisis, the small amount of inventory in the business can be a risk. In other words, the organization is heavily dependent on the manufacturers and the workers [13].

Along with these challenges, there are some environmental challenges also. You will have to note that there will be some ethical and moral concerns from the workers as you work and introduce lean in a working process, because the introduction of lean will directly impact their everyday work process [13]. Discussions can lead to disagreements and uncertainty, so the lean framework for the business needs to be articulated in a well-structured, evaluated and positive manner. The benefits can help workers to move toward a leaner way of working (especially in numbers). The persuasion that lean manufacturing should be introduced would need to be explicitly and substantially articulated for the best possible reaction and acceptance by the workers [13]. The transition to lean production requires a major change not only for the workers, but for the company's whole culture, which can be difficult and costly to create. There is a possibility that there will be more waste than profit development in the organization if the company leaders lack persuasive skills to solve the social challenges. The most difficult part of the transmission may be that the workers would not acknowledge the improvements made, even though they have been provided with obvious beneficial aspects. This may lead to tension because workers can feel that they are too disciplined or that the decision-makers are looking for a lack of competence [13]. There are some environmental implications of lean production that are necessary to consider when leaning into a business is introduced. As companies pursue lean manufacturing, they tend to seek methods that can minimize the materials, resources, water, space, etc. But what is not directly targeted are the environmental endpoints posed by lean production processes. This includes, for example, radioactive waste, air pollution and the discharge of wastewater [13]. When implementing lean production, these forms of environmental effects are challenging to take into account since the lean approaches are typically restricted to a subsystem, which is primarily the company's operating process, and not external influences [13].

It is difficult to find accurate information from only one discussion in relation to discrepancies or modifications in Lean related to the development of smart and connected goods. To collect primary data, a clear observation of the manufacturing process as well as of applied Lean methods would be more suitable. That was not possible due to limited time and access to data on the case-company side [1]. The variations or adjustments in production are becoming more electronic components and software because the goods are getting smarter and related. Because of that, more content needs to be sourced from outside. The electronic parts are highly sensitive and must be shielded from electrostatic discharge (ESD). In addition, new operating equipment is required, such as testing, and people need new information about new types of products and processes. For that, to gain more abilities, they have to be qualified. However, because of technical advancement and further outsourcing, certain skills are no longer required. For each, the degree of transition is not the same. It depends on the role within the organization. Cooperation between the departments must be closer in certain cases [1]. There are variations or improvements associated with the Lean system, too. New forms of waste happen. There are new software issues that lead to reworking and have a negative effect on lead time and process flow. Employees, particularly older people, may have problems and feel anxious about the move toward digitization. The way that Lean thinks is still the same. Nevertheless, due to a lack of know-how and/or familiarity with the software, the use of Lean tools may be more complex, complicated and time-consuming.

It may be more difficult for people to come up with changes related to the Kaizen process. In addition, the upgrades themselves are becoming more digital [1]. Finally, outside of the manufacturing department, there are improvements as well. New information about the electronics industry is required throughout the sourcing phase.

Adoption of Industry 4.0 in Lean Manufacturing

Finding new, trustworthy suppliers is crucial and new people have to be recruited. Because of this, new business areas must be developed. Many new employees in the R&D department must also be recruited because of software development [1].

6.19 CONCLUSION

Industry 4.0 provides huge opportunities for revolutionary companies, entire markets, and service providers. Still much like previous technical innovations, Industry 4.0 is nonetheless a major barrier for the laggards. As market dynamics, demands, and economies for abilities shift, we might well see extreme changes in top positions in both regional and organizational levels. Fabrication methods are able to improve versatility and permit affordable advancement of small batch sizes. Robots, smart devices and smart machines which speak with one another and make specific autonomous decisions will give this versatility. The production processes could be made simple across the supply chain by using automated IT methods. So, today's insular production cells are likely to be replaced with fully integrated, interconnected production lines. Equipment, production methods, and automation of production are intended and implemented in basically one integrated procedure and often by supplier–producer partnership. They are going to reduce real physical prototypes to a complete minimum. Manufacturing processes can be improved by both knowing and self-optimizing gear cycles that can, as an example, alter their own parameters whenever they feel the incomplete item's attributes. The theories suit, because each Lean and Industry 4.0.0. are influenced by productivity. But the primary Lean function installed on the platform was predictive maintenance, as the technology is still in the early stages. Another possible point of view that has been often proposed is Assets Performance Management (APM). The idea also includes predictive maintenance, which may provide more insights into how the company is monitoring its assets and how the trend toward more real-time data is evolving. One has to understand that APM is not a theoretical discipline but instead a realistic term. Nevertheless, this field seems to be evolving strongly from a market perspective.

REFERENCES

1. Raymann, R. (2018). *Digitisation & lean manufacturing: changes in manufacturing when the products are getting smarter and connected*, Dissertation, University of Gävle.
2. Leitão, P., Rodrigues, N., Barbosa, J., Turrin, C., and Pagani, A. (2015). Intelligent products. The grace experience. *Control Engineering Practice*, Vol. 42, pp. 95–105.
3. Hoellthaler, G., Braunreuther, S., and Reinhart, G. (2018), Digital lean production. an approach to identify potentials for the migration to a digitalized production system in SMEs from a lean perspective. *Procedia CIRP*, Vol. 67, pp. 522–527.
4. Comstock, B. (2016). *Minds & machines*. San Francisco, General Electric.
5. Manyika, J. et al. (2015). The internet of things: mapping the value beyond the hype, s.l.: McKinsey&Company.
6. General Electric (2016a). *Minds & machines*. [Online] Available at: https://www.ge.com/digital/minds-machines [Accessed 29 August 2020].
7. Womack, J., Jones, D., and Roos, D. (1990). *The machine that changed the world*. 1st ed. New York, NY, Rawson Associates.

8. De Treville, S. and Antonakis, J. (2006). Could lean production job design be intrinsically motivating? Contextual, configurational, and levels-of-analysis issues. *Journal of Operations Management*, Vol. 24, pp. 99–123.

9. Narasimhan, R., Swink, M., and Kim, S. W. (2006). Defining lean production: some conceptual and practical issues. *Journal of Operations Management*, Vol. 24, pp. 440–457.

10. Moyano-Fuentes, J. and Sacristan-Diaz, M. (2012). Learning on lean: a review of thinking and research. *International Journal of Operations & Production Management*, Vol. 32, No. 5, pp. 551–582.

11. Olsson, J. G. and Xu, Y. (2018). *Industry 4.0 adoption in the manufacturing process: multiple case study of electronic manufacturers and machine manufacturers*, Thesis, Linnaeus University.

12. Sundar, R., Balaji, A. N., and SatheeshKumar, R. M. (2014). A review on lean manufacturing implementation techniques. *Procedia Engineering*, Vol. 94, pp. 1875–1885.

13. Ghanem, M. (2020). *Manufacturing process re-engineering of a production line through Industry 4.0 to obtain the best quality and reduced wastes: the case in projection welding*, Thesis, Blekinge Institute of Technology.

14. Industry 4.0: the impact of horizontal integration on manufacturing business models and intellectual property strategies Kauffman, M. (Author), 2020.

15. Gutiérrez Gutiérrez, L. J., Lloréns-Montes, F. J., and Bustinza Sánchez, Ó. F. (2009). Six sigma: from a goal-theoretic perspective to shared-vision development. *International Journal of Operations and Production Management*, Vol. 29, No. 2, pp. 151–169.

16. Dahlgaard, J. J. and Dahlgaard-Park, S. (2005). 'TQM and company culture'. industrial management data systems. *International Journal of Operations and Production Management*, Vol. 18, No. 7, pp. 263–281.

17. Lasi, H., Fettke, P., Kemper, H. G., Feld, T., and Hoffmann, M. (2014). Industry 4.0. *Business and Information Systems Engineering*, Vol. 6, No. 4, pp. 239–242.

18. Caputo, A., Marzi, G., and Pellegrini, M. M. (2016). The Internet of Things in manufacturing innovation processes: development and application of a conceptual framework. *Business Process Management Journal*, Vol. 22, No. 2, pp. 383–402. doi:10.1108/BPMJ-05-2015-0072

19. Roblek, V., Mesko, M., and Krapez, A. (2016). *A complex view of industry 4.0*. London, Sage Publications.

20. Agerskans, N., (2020). *A Framework for Achieving Data-Driven Decision Making in Production Development*, Dissertation.

21. Chen, H., Chiang, R. H. L., and Storey, V. C. (2012). Business intelligence and analytics: from big data to big impact. *MIS Quarterly*, Vol. 36, No. 4, pp. 1165–1188.

22. Fayyad, U., Piatetsky-Shapiro, G., and Smyth, P. (1996). From data mining to knowledge discovery in databases. *AI Magazine*, Vol. 17, No. 3, pp. 37–54.

23. Arnott, D. and Pervan, G. (2008). Eight key issues for the decision support systems discipline. *Decision Support Systems*, Vol. 44, No. 3, pp. 657–672.

24. Kahneman, D. (2003). Maps of bounded rationality: psychology for behavioral economics. *American Economic Review*, Vol. 93, No. 5, pp. 1449–1475.

25. Ghobakhloo, M. and Fathi, M. (2020). Corporate survival in Industry 4.0 era: the enabling role of lean-digitized manufacturing. *Journal of Manufacturing Technology Management*, Vol. 31, No. 1, pp. 1–30. doi:10.1108/JMTM-11-2018-0417

26. Pehrsson, A. (2020). *Industry 4.0: impact on manufacturing strategies and performance*, Dissertation.

27. Hosseinzadeh Dadash, A. (2020). *A cyber-physical testbed for wireless networked control systems*, Dissertation.

Adoption of Industry 4.0 in Lean Manufacturing

28. Lee, E. A. *Cyber physical systems: design challenges*. In: *11th IEEE Symposium on Object Oriented Real-Time Distributed Computing (ISORC)*, Orlando, FL, USA.
29. Rajkumar, R., Lee, I., Sha, L., and Stankovic, J. (2010). *Cyber-physical systems: the next computing revolution*. In: *Design Automation Conference*, Anaheim, CA, pp. 731–736.
30. Frontoni E., Loncarski J., Pierdicca R., Bernardini M., and Sasso M. (2018). Cyber physical systems for industry 4.0: towards real time virtual reality in smart manufacturing. In: De Paolis L., Bourdot P. (eds) *Augmented Reality, Virtual Reality, and Computer Graphics*. AVR. Lecture Notes in Computer Science, Vol. 10851. Cham, Springer.
31. Lee, J., Bagheri, B., and Kao, H.-A. (2015). A cyber-physical systems architecture for industry 4.0-based manufacturing systems. *Manufacturing Letters*, Vol. 3, pp. 18–23.
32. Wang, E. K., Ye, Y., Xu, X., Yiu, S. M., Hui, L. C. K., and Chow, K. P. (2010). *Security issues and challenges for cyber physical system*. In: *Proceedings of the 2010 IEEE/ACM International Conference on Green Computing and Communications & International Conference on Cyber, Physical and Social Computing* (pp. 733–738). Hangzhou, China, IEEE Computer Society.
33. He, K. and Jin, M. (2016). *Cyber-Physical systems for maintenance in Industry 4.0*, Dissertation.
34. MacDougall, W. (2014). *Industrie 4.0: smart manufacturing for the uture*. Berlin, Germany, Germany Trade & Invest.
35. Chen, M., Jin, H., Wen, Y., and Leung, V. (2013). Enabling technologies for future data center networking: a primer. *IEEE Network*, Vol. 27, No. 4, pp. 8–15.
36. Lee, J., Bagheri, B., and Kao, H. A. (2014a). *Recent advances and trends of cyber-physical systems and big data analytics in industrial informatics*. In: *International Conference on Industrial Informatics (INDIN)*, Porto Alegre, Brazil.
37. Wang, S., Wan, J., Li, D., and Zhang, C. (2016). Implementing smart factory of industries 4.0: an outlook. *International Journal of Distributed Sensor Networks*, Vol. 12, No. 1, pp. 1–10.
38. Jain, A., Bhatti, R., and Singh, H. (2014), Total productive maintenance (TPM) implementation practice: a literature review and directions. *International Journal of Lean Six Sigma*, Vol. 5, No. 3, pp. 293–323.
39. De Vasconcelos Batalha, A. and Parli, A. L. (2017). *Industry 4.0 with a Lean perspective - Investigating IIoT platforms' possible influences on data driven Lean*, Dissertation.
40. Atzori, L., Iera, A., and Morabito, G. (2010). The Internet of Things: a survey. *Computer Networks*, Vol. 54, pp. 2787–2805.
41. Herrero Cuesta, J. M. and Molina Gil B. (2019). '*Study of a lean manufacturing laboratory: improvements towards automation*', Dissertation.
42. Woman, J. P. and Jones, D. T. (1996). Lean thinking - banish waste and create wealth in you corporation. *Journal of the Operational Research Society*, Vol. 48, No. 11, pp. 1148–1148.
43. Liker, J. K. (2004). *The Toyota way: 14 management principles from the World's Greatest Manufacturer*. New York, McGraw-Hill.
44. Larsson, J. and Wollin, J. (2020). *Industry 4.0 and lean–possibilities, challenges and risk for continuous improvement: an explorative study of success factors for industry 4.0 implementation*, Thesis, Blekinge Institute of Technology.
45. Mrugalska, B. and Wyrwicka, M. K. (2017). Towards lean production in industry 4.0. *Procedia Engineering*, Vol. 182, pp. 466–473.

7 Internet of Things-Based Economical Smart Home Automation System

Pawandeep Kaur and Krishan Arora

Lovely Professional University, Phagwara, India

CONTENTS

7.1 Introduction 129
7.2 Pros of Home Automation Systems (HAS)..130
7.3 Existing System...131
7.4 Research Gaps 133
7.5 Proposed System 134
7.6 Methodology 134
    7.6.1 Framework 134
    7.6.2 Component Requirement Analysis 135
        7.6.2.1 Hardware Requirements...135
        7.6.2.2 Software Requirements 135
7.7 Implementation 136
7.8 Functional Requirements 136
7.9 Results and Discussion..138
7.10 Conclusion...141
References..141

7.1 INTRODUCTION

Automation is evolving in every field around us. As person living in the fastest growing age of technology development, we have moved from room-sized computers to light and portable laptops, from insecure databases to highly secure web. Same is the case for homes. There are several technological developments in this field of automation and one of them is automation of home or "Smart Homes". By connecting individual devices and home appliances to a centralized hub and controlling each and every thing through that network is quite fascinating. Smart home automation allows you to experience high-tech functionality and convenience that wasn't possible in the past.

Numerous applications on IoT in different fields are medical as e-health, smart grid, smart cities, smart education system, smart transportation, smart agriculture, smart factories, etc. Above all field applications smart home (SH) has major attraction in academic and industrial due to its direct relation with the real life [1–3]. The previous works on smart technologies have focused on saving time and providing

DOI: 10.1201/9781003102267-7

129

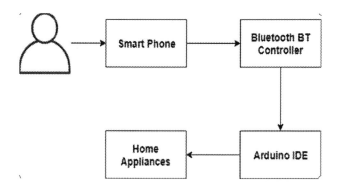

FIGURE 7.1 Bluetooth-based automation system.

comfort to elderly and disable person. Intelligence is a way of controlling home planting or gardening along with the security and surveillance. As per literature [4, 5] Figure 7.1 depicts the various electronic gadgets with communication interface based on wireless, Bluetooth, and Wi-Fi.

Home automation system (HAS) is available in two broad categories based on controlled methodology as (i) native controlled and (ii) remotely controlled [6]. The former controlled system depicts that user can control home appliances by using controllers in the form of wireless (ZigBee, GSM, Bluetooth) or stationary communication methods, whereas in the case of remotely control users can control their home appliances with Internet connections through mobile phones. There are numerous flaws while setting up the controlling procedure; thus, the requirement is that HAS should provide understanding interface for observing and controlling the appliances [6].

7.2 PROS OF HOME AUTOMATION SYSTEMS (HAS)

Nowdays, for fast and reliable connection wireless communication systems like Wi-Fi become more promising candidate with ample advantages over wired connections as follows:

  i. Cost effective: As the installation cost is decreased due to nonrequirement of cables as compared with wired system in which the cable, its material, and laying technology along with professionals added cost.
  ii. Easy deployment, installation, and coverage: Remote devices or appliances can be easily accessed without wire along with easy deployment and installation of wireless system.
  iii. System extension and scalability: Transport of a native structure is predominantly priceless when, due to novel or distorted prerequisites, augmentation of the scheme is essential. Somewhat than wired establishments, in which cabling extension is expensive and time-consuming [7]. This makes remote systems as an original essence.
  iv. Amalgamation of mobile devices: With advancement of the communication technology wireless platform gives the automation an easier and possible solution, where remotely devices or appliances can control or observe without any physical connections.

v. Aesthetical benefits: Iit includes the the architecture with clear glimpses of historical buildings, monuments and tourist spots.
vi. Security and aid for old age: This home automation not only comforts the teenagers but also it is beneficial for physically challenged and aged person [8]. For all above reasons, the communication or data transfer technology is choice of innovation, renovate, and new installations.

7.3 EXISTING SYSTEM

Currently if we look at other home automations in the market, there are a lot of techniques they use. Different modes of wireless communication used to transfer of data and signals between their various components such as Bluetooth, ZigBee, wireless-fidelity, and global system for mobile (GSM). Bluetooth (HC-05) acts as a slave device that is used for communication among Arduino (Mega) IDE and smartphone. The operating voltage is 3.6– 6 V. Bluetooth Board HC-06 has six pins such as State, RXD, TXD, GND, VC C, and EN. Figure 7.1 shows the block diagram illustrating the connection mode of home appliances with the smartphone through Bluetooth BT cell with enhanced data rate at 3 Mbps modulation at frequency of 2.4 GHz. Internally input/output port of Arduino board (microcontroller) is programmed using C language/Python and the connection made to home appliances through Bluetooth BT cell [9].

ZigBee is under radio frequency (RF) communication, having an IEEE 802.15.4 standard. The reflection of the ZigBee standard illustrates the packet routing between the network in which a large number of nodes and transmission of the data are in zig-zag pattern as shown in Figure 7.2. ZigBee supports application in all fields. The main component in ZigBee is its coordinator (ZC) that maintains the network and routing table. This standard has three frequency bands, and prominent band use is at 2.5 GHz at data rate of 250 and 40 Kbps. This method has numerous advantages as per literature reported as low cost, inter-operatability, low signal-to-noise ratio, low power consumption, implemented by using any microcontroller, innovations are easily to deploy, high data security, and low bandwidth required for minor projects [10] (Table 7.1).

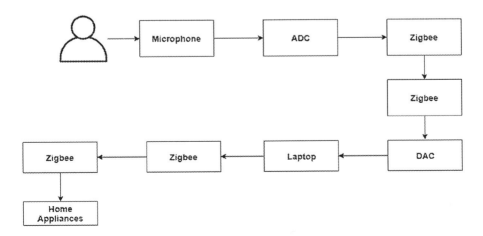

FIGURE 7.2 ZigBee home automation system.

TABLE 7.1
Comparison Between Different Methodologies

Methodologies	Communication Interface	Controller	User Interface	No. of Devices	Cost	Speed	Range	Energy Saving
Bluetooth	Bluetooth and AT instructions	Ardunio-uno	Android	Unlimited	Moderate	Low	10 m	No
GSM	SMS messages	Arduino ADK	Mobile phone	Limited	Low	Very low		No
ZigBee	ZigBee with Instructions	ZigBee Controller (ZC)	Smartphone	Unlimited	High	Low	10 m	Yes
Wireless Proposed Design	Wi-Fi (NODEMCU)	Arduino SDK	Smartphone	Unlimited	Low	High	unlimited	Yes

Internet of Things-Based Economical Smart Home Automation System

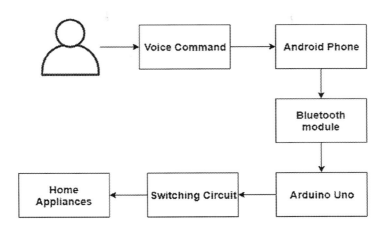

FIGURE 7.3 Voice recognition.

As per the literature, voice control method is used for home automation as shown in Figure 7.3 with the help of Arduino, various wireless technologies such as GSM, Bluetooth, Wi-Fi, and NFC. In [11–13], authors have performed their experiment on five groups. Each group comprises of different voice commands on Access, fan, light, utility, and the safety were performed at baud rate of 115200 bps to make smart home automation system (SHAS) [14]. The advanced feature along with voice is used for recognition of these systems [15]. The home automation is controlled through voice by using Arduino as controller and the Bluetooth HC-05 module as wireless communicator and HC-SR04 as an ultrasonic sensor to detect the motion as shown in Figure 7.3.

7.4 RESEARCH GAPS

Technology is advancing in every field and that's how the concept of home automation came. Technology is all about making each and every task easier. There are different technologies that are used to automate your home or offices that are discussed. These implementations have some problem or something that can be developed like:

i. **Dependency on range**: HASs that connect all the devices to the mobile app through Bluetooth works when you are in the range of the central hub. Generally, the range for Bluetooth connectivity is 100 m [16]. You cannot control your home appliances from outside of this range.

ii. **Expandability:** New home assistants like Alexa and Google assistants are trending a lot these days but the appliances they interact with are limited. They can only interact with smart appliances and won't control our standard home appliances like tube light, fan, or old LCD TV [16, 17]. The other way to control them is hardwiring. That again is an expensive alternative.

iii. **Economical:** The main motive of technology is to empower each and every individual but sometimes the cost of a new technology is a barrier [18]. Being new in the market, home automations are quite expensive and cannot be afforded by everyone.
iv. **Energy Saving**: In this regard, the saving of the electricity is automatically switched off the appliances when not required. Similarly, nowadays smart grid system [19] helps to save energy with the implementation of smart condenser banks that cover all reactive power generated by the current transformers (CT) and potential transformers (PT). Monitoring the electricity usage can be helpful in saving the energy and predicting the cost for an individual.

7.5 PROPOSED SYSTEM

In the proposed system, there are a total of three modules; each module has been made for different purposes and each of them functions differently. The mobile application is to provide a user-friendly interface that will help to organize and observe the usage of home appliances. For this module, the enhanced features of Android and Java programming language are used. The second purpose is to connect the application to a database; this can be achieved through firebase. The third module in the system is to connect devices to a central hub that is then sending and receiving data from the server. This feature has been done using IOT devices and C programming languages.

7.6 METHODOLOGY

7.6.1 FRAMEWORK

In the above design flow after analyzing the functionality and feasibility of the problem, the requirements of hardware and software for the communication are framed as explored in the below section. The software section consists of design of mobile application with the login and registration page having the designed home page and dedicated page for each device. In addition, more functionality to the application like monitoring energy usage, controlling devices, etc., is added as shown in Figure 7.4.

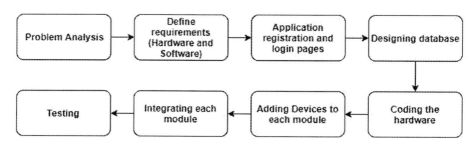

FIGURE 7.4 Design flow of proposed system.

Internet of Things-Based Economical Smart Home Automation System   135

Then the designing of the real-time database to store the state of each device in the system is performed. Code is written for hardware components and connected them to the database created earlier along with the application. After this, an authentication and authorization page was created that enabled login through Google and connected different home appliances to our central hubs like fan, bulb, and mobile charger. Finally, the j-unit-testing was done for modules. The manual application testing on different android devices running different platforms from Lollipop to Android 10 is explored [19].

7.6.2 Component Requirement Analysis

The major categorization of the projected system is in two hardware and software requirements. Thus in this segment, the analysis of the hardware components and software is explored. The features and demanded configuration of the microcontroller are studied along with the compatible softwares.

7.6.2.1 Hardware Requirements
 I. Relay (an electrically operated switch)
 II. Wi-Fi module (ESP8266 NODEMCU) (integrated TR switch, low-noise amplifier, TCP/IP protocol, integrated PLL, regulator and power organization section, along with 19.5 dBm output power in 802.11b mode as shown in Figure 7.5.
 III. Jumper wires
 IV. General-purpose PCB
 V. Temperature and humidity sensor (DHT11) (digital type, capacitive humidity sensor, it uses thermistor for measuring outside air, and send digital signal on data pins)

7.6.2.2 Software Requirements
 I. Android Studio 3.0.1
 II. Python 2.7
 III. Arduino IDE 1.8.9

FIGURE 7.5   Wi-Fi module (ESP8266 NODEMCU).

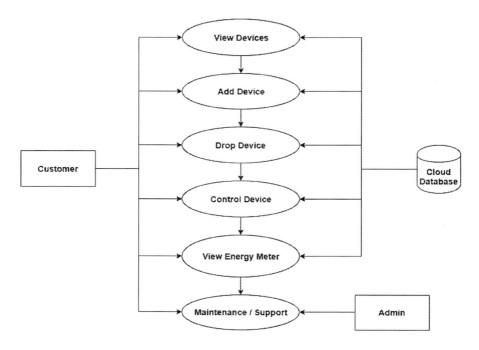

FIGURE 7.6 Functionalities in HAS application.

7.7 IMPLEMENTATION

In the following section, the flow graph depicting the user can control and observe the functionalities of the developed application that incorporates the devices (visualization, addition, deleting, and controlling) along with the reading about the energy consumption feature that is newly added in the SHAS as shown in Figure 7.6. Moreover, the maintenance or support feature is controlled by the administrator in addition to customers.

All related information is shared in a cloud database that runs on a cloud computing platform, in which the data can be accessed as-a-service provider. In database services, scalability and high availability of the database is ensured by the database service provider. Figure 7.7 indicate various services offered by Google specially storage, real-time database, online processing, authorization of the user along with their updated usage information.

7.8 FUNCTIONAL REQUIREMENTS

Functional Requirements 1: Measurement of parameters estimating the considerable number of parameters, for example, current stream, vitality utilization, and time interim for which gadget is being utilized in room is

Internet of Things-Based Economical Smart Home Automation System

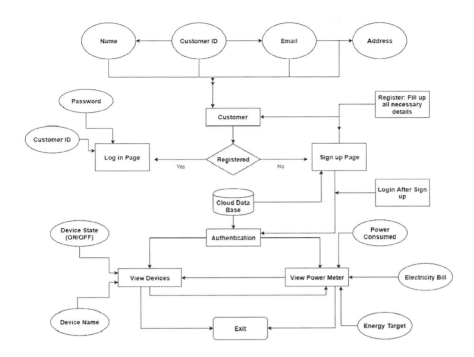

FIGURE 7.7 Implementation of proposed architecture of application.

important to spare and control the gadgets by utilizing the different sensors and versatile application.

Functional Requirements 2: Fetching of data sensors measures the parameters and those parameters fetch for further process like data analysis. Measuring and fetching data are the basic functional requirements of this system.

Functional Requirements 3: Manipulation of data after fetching data and manipulation is done with provided database and the current data. They are analyzed to generate the ideal solution.

Functional Requirements 4: Device selection device selection process looks for the time interval and gives the notification that is used to turn off the device.

Functional Requirements 5: Notification on mobile phone while detecting if there is any device being used for more than its usual time; it notifies the user that there is going something wrong.

Functional Requirements 6: Registration/login users can login into the system; they can get usage time, notification into mobile phone, can give feedback for any sort of improvement, and can complain about error. Admin can manage the entire database and the system and manage the complaining from farmer and can give best suggestions to them.

7.9 RESULTS AND DISCUSSION

In the following section, the consequences are based on the accomplishment of the application as proposed and various images shown in this section as login page, registration, and home and control page. The login page shown in Figure 7.8 provides the option for the user to login with their credentials into the app. For a first-time installation, the user must register for the app on the registration page shown in Figure 7.9. In this page, a user can set up his profile and by entering an e-mail id and password for the further usage of the app.

FIGURE 7.8   Login page.

Internet of Things-Based Economical Smart Home Automation System 139

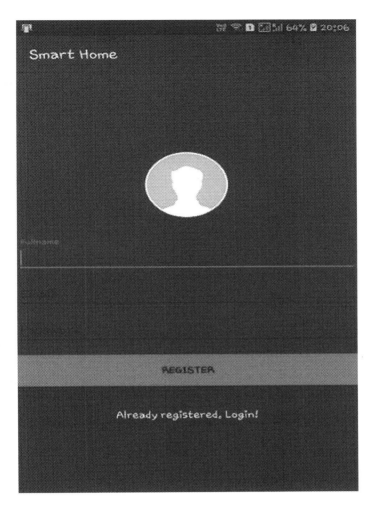

FIGURE 7.9 Registration page.

Also, they'll have to confirm their mail id by clicking on the link sent to their email id. The purpose of email id is that if in any case a user forgets the password then it can be retrieved using the email id shown in Figure 7.10. Appliances can be switch on/off using the Home page. After logging in to the application there will be a list of devices available. Devices can be renamed according to the appliances connected to them. Power consumption of each appliance is shown on the home page shown in Figure 7.11.

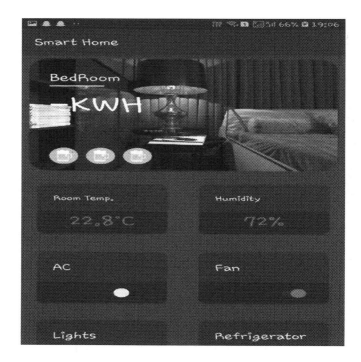

FIGURE 7.10 Home page.

FIGURE 7.11 Control page.

7.10 CONCLUSION

In the nutshell, the implemented app is working successfully with hardware as an Arduino controller. The advanced feature is energy consumption reading shown in the app; thus, the user can control the saving as per requirement along with the addition and deletion of the devices. This chapter also supports the real-time database, online processing, and authentication and authorization of the user by means of implemented application names as "Smart Home".

REFERENCES

[1] W. A. Jabbar, M. Ismail, and R. Nordin, *"Evaluation of energy consumption in multipath OLSR routing in Smart City applications,"* In: *2013 IEEE 11th Malaysia International Conference on Communications (MICC)*, Malaysia, pp. 401–406, 2013.

[2] K.-M. Lee, W.-G. Teng, and T.-W. Hou, "Point-n-Press: An intelligent universal remote control system for home appliances," *IEEE Transactions on Automation Science and Engineering*, vol. 13, pp. 1308–1317, 2016.

[3] P. P. Gaikwad, J. P. Gabhane, and S. S. Golait, *"A survey based on Smart Homes system using Internet-of-Things,"* In: *2015 International Conference on Computation of Power, Energy, Information and Communication (ICCPEIC)*, Melmaruvathur, Chennai, India, pp. 0330–0335, 2015.

[4] W. A. Jabbar, M. Ismail, and R. Nordin, *"MBA-OLSR: a multipath battery aware routing protocol for MANETs,"* In: *2014 5th International Conference on Intelligent Systems, Modelling and Simulation*, Langkawi, Malaysia, pp. 630–635, 2014.

[5] T. Song, R. Li, B. Mei, J. Yu, X. Xing, and X. Cheng, "A privacy preserving communication protocol for IoT applications in smart homes," *IEEE Internet of Things Journal*, vol. 4, pp. 1844–1852, 2017.

[6] R. Piyare and M. Tazil, *"Bluetooth based home automation system using cell phone,"* In: *2011 IEEE 15th International Symposium on Consumer Electronics (ISCE)*, Singapore, pp. 192–195, 2011.

[7] W. A. Jabbar, M. Ismail, R. Nordin, and S. Arif, "Power-efficient routing schemes for MANETs: a survey and open issues," *Wireless Networks*, vol. 23, pp. 1917–1952, 2017.

[8] M. N. Jivani, "GSM based home automation system using app-inventor for android mobile phone," *International Journal of Advanced Research in Electrical, Electronics and Instrumentation Engineering*, vol. 3, 2014.

[9] E. M. Wong, "A phone-based remote controller for home and office automation," *IEEE Transactions on Consumer Electronics*, vol. 40, pp. 28–34, 1994.

[10] K. Gill, S.-H. Yang, F. Yao, and X. Lu, "A ZigBee -based home automation system," *IEEE Transactions on Consumer Electronics*, vol. 55, pp. 422–430, 2009.

[11] M. R. Kamarudin, M. Yusof, and H. T. Jaya, "Low cost smart home automation via microsoft speech recognition," *International Journal of Engineering & Computer Science*, vol. 13, pp. 6–11, 2013.

[12] B. Yuksekkaya, A. A. Kayalar, M. B. Tosun, M. K. Ozcan, and A. Z. Alkar, "A GSM, internet and speech controlled wireless interactive home automation system," *IEEE Transactions on Consumer Electronics*, vol. 52, pp. 837–843, 2006.

[13] S. Kumar and S. R. Lee, *"Android based smart home system with control via Bluetooth and internet connectivity,"* in *The 18th IEEE International Symposium on Consumer Electronics (ISCE 2014)*, Jeju, South Korea, pp. 1–2, 2014.

[14] Y. Mittal, P. Toshniwal, S. Sharma, D. Singhal, R. Gupta, and V. K. Mittal, *"A voice-controlled multi-functional smart home automation system,"* In: *2015 Annual IEEE India Conference (INDICON)*, New Delhi, India, pp. 1–6, 2015.

[15] M. E. Abidi, A. L. Asnawi, N. F. Azmin, A. Jusoh, S. N. Ibrahim, H. A. M. Ramli, et al. *"Development of voice control and home security for smart home automation,"* In: *2018 7th International Conference on Computer and Communication Engineering (ICCCE),* Kuala Lumpur, Malaysia, pp. 1–6, 2018.

[16] A. Z. Alkar and U. Buhur, "An Internet based wireless home automation system for multifunctional devices," *IEEE Transactions on Consumer Electronics,* vol. 51, pp. 1169–1174, 2005.

[17] N. Dickey, D. Banks, and S. Sukittanon, *"Home automation using Cloud Network and mobile devices,"* In: *2012 Proceedings of IEEE Southeastcon,* Florida, USA, pp. 1–4, 2012.

[18] Y. Cui, M. Kim, Y. Gu, J.-J. Jung, and H. Lee, "Home appliance management system for monitoring digitized devices using cloud computing technology in ubiquitous sensor network environment," *International Journal of Distributed Sensor Networks,* vol. 10, p. 174097, 2014.

[19] Q. Yang, "Internet of things application in smart grid: a brief overview of challenges, opportunities, and future trends," In: *Smart Power Distribution Systems,* Manhattan, New York, Elsevier, pp. 267–283, 2019.

8 Machine Vision Technology, Deep Learning, and IoT in Agricultural Industry

K. Harjeet and Deepak Prashar

Lovely Professional University, Phagwara, India

CONTENTS

8.1 Introduction to Smart Farmin 144
    8.1.1 Crop Management 145
    8.1.2 Field Condition Management 145
    8.1.3 Livestock Management 145
    8.1.4 Pest Management 145
    8.1.5 Weather Forecasting 146
8.2 Machine Vision Technology 146
    8.2.1 Components of Machine Vision 147
    8.2.2 Machine Vision in Agriculture 148
8.3 Deep Learning and Its Techniques 148
    8.3.1 Feed-Forward Neural Network and Back Propagation (BP) 148
    8.3.2 Convolutional Neural Networks (CNN/ConvNet) 148
    8.3.3 Recurrent Neural Networks (RNN) 148
    8.3.4 Generative Adversarial Networks (GAN) 149
8.4 Advantages of Combining Machine Vision and Deep Learning 150
8.5 IoT Solutions to Agricultural Problems 150
    8.5.1 Advantages of IoT in Smart Farming 151
    8.5.2 IoT Use Cases in Smart Farming 151
        8.5.2.1 Climate Conditions Monitoring 151
        8.5.2.2 Automation of Greenhouse 152
        8.5.2.3 Management of Crops 152
        8.5.2.4 Cattle Monitoring and Management 152
        8.5.2.5 Precision Farming 152
        8.5.2.6 Agricultural Drones 153
        8.5.2.7 Predictive Analytics for Smart Farming 153

DOI: 10.1201/9781003102267-8

	8.5.3	Challenges in IoT Based Smart Farming	153
		8.5.3.1 Sensor Selection	153
		8.5.3.2 Appropriate Data Analytical Tools	153
		8.5.3.3 Maintenance of IoT Setup	153
		8.5.3.4 Connectivity	154
		8.5.3.5 Data Security	154
		8.5.3.6 Data Collection Frequency	154
8.6	Agrobots – Agricultural Robots		154
	8.6.1	Agrobots Working in Fields for Different Agricultural Activities	155
	8.6.2	Harvesting Robots	157
	8.6.3	Challenges in Implementing Agrobots	157
	8.6.4	Conclusion and Future Work	158
References			158

8.1 INTRODUCTION TO SMART FARMING

Agriculture has a significant impact on the world's economy. With the continuous expansion of human population, pressure on agricultural system is increasing to meet the requirements. In today's world also, there are many countries that rely on old, traditional ways of farming that sometimes leads to insufficient production of grains or low-quality crops. Precise, accurate, and fast decision-making systems are required for better fieldwork which can be achieved with the use of modern machinery and automated robots in agricultural fields. Many researchers are working on combining technology with farming to improve productivity with minimum adverse effects on environment. Problems associated with conventional agriculture system and different approaches to solve them, from databases to decision support systems, are suggested in different works [1–3]. Systems those make use of Artificial Intelligence (AI) are becoming more popular choice because of evident accuracy and robustness. AI techniques can solve complex problems with better accuracy and preciseness than other techniques.

The Industrial Internet of Things (IIoT) has impacted many industries and the Agriculture Industry isn't an exception. Smart farming is an emerging concept involving the use of technologies like IoT, robotics, drones and AI to increase the agricultural productivity in terms of both quality and quantity with minimized human involvement. Idea of Smart Farming revolves around the use of technology in agricultural fields with clear cut objectives of improved quality and quantity of crops. Smart farming is application of AI techniques and machine learning (ML) algorithms both in farm products and in field farming techniques [4, 5]. To further enhance the performance, deep learning (DL) algorithms may be applied in different aspects of agriculture. Given below, and as shown in Figure 8.1, are five major areas where the use of AI solutions can benefit agriculture or farming.

- Crop Management
- Field Condition Management
- Pest Management

Machine Vision Technology, Deep Learning, and IoT in Agricultural Industry 145

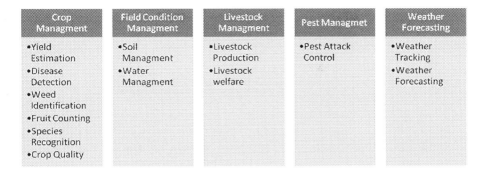

FIGURE 8.1 Areas for implementing AI solutions in smart farming.

- Livestock Management and
- Weather Forecasting

8.1.1 CROP MANAGEMENT

Crop management includes crop yield estimation, fruit counting, crop disease detection, weed identification, species recognition, and crop quality. Estimation of cost is significant for every stakeholder including farmers, consumers, and government. Complete strategies for selling, purchasing, and marketing are dependent on this. Fruit counting is used in robotic harvesting and is also very significant as far as yield prediction is concerned. Manually working on these subareas is tedious as well as time-consuming. Application of AI and DL in crop management are giving accurate results. Leaf image classification and pattern recognition are used in plant disease identification.

8.1.2 FIELD CONDITION MANAGEMENT

Field condition management includes soil and water management. AI and DL when applied in fields can be used to detect defects and deficiency in soil. Soil health monitoring is very crucial for good crops. Smart irrigation with more effective use of water resources is also utilizing AI techniques.

8.1.3 LIVESTOCK MANAGEMENT

Livestock management is categorized as livestock production and animal well-being. Smart farming techniques are helpful in monitoring the needs and health of individual animals. Consequently, their diet can be adjusted to make it more nutrient that protects them from unwanted diseases and enhances herd health.

8.1.4 PEST MANAGEMENT

Pests are always haunting farmers as crop gulping insects like grasshoppers, locusts eat profits of farmers. But AI gives growers a weapon against cereal-hungry bugs.

146 Industrial Internet of Things

Farmers make wrong decisions in choosing pesticides and fertilizers for pest control. But with the application of precision agriculture, variation within the field can be measured and farmers can change strategies accordingly. It results in better pest management.

8.1.5 WEATHER FORECASTING

Harvesting process heavily depends upon weather conditions and rains. Continuous tracking of weather condition is required especially at the time of crop cutting. Farmers are making use of their own experience and intuitions to guess the mood of weather. DL applications in this field is assisting farmers.

For all the applications mentioned above, appropriate usage of IOT, drones, sensors, machine vision technology, and ML algorithms are required for smart farming.

8.2 MACHINE VISION TECHNOLOGY

Machine vision technology is used to provide automation of image-based inspection and analysis in applications, both industrial and non-industrial, which uses automatic inspections or process. The major use of machine vision is in robot supervision. Applications of machine vision combines hardware and software to give operational guidance to robots functioning on the bases of captured images. Machine vision systems rely on digital sensors protected inside industrial cameras with specialized optics to acquire images, so that computer hardware and software can process, analyze, and measure various characteristics for decision-making. Industrial applications of machine vision needs more reliability, robustness, and stability but low cost as compared to non-industrial applications. Fill-level inspection system at a brewery is an industrial example of machine vision technology as shown in Figure 8.2. Each bottle of bear is passed through an inspection sensor and picture of the bottle is taken. Then image is processed and then analyzed for appropriate fill level. Apart from industry, applications of machine vision are apparent in the field of medical, signature identification, optical character recognition, hand writing recognition, object or pattern recognition, electronic component analysis, material inspection, currency inspection, etc. The first step in every machine vision application is pattern matching for identification of features or object of interest from the images collected through camera. Identification may be successful or unsuccessful. Although vision systems are trained to recognize features or patterns still it is challenging to identify them correctly and precisely. To get accurate, precise and reliable results vision system's pattern recognition system must include sufficient intelligence. Four major applications of machine vision include – Guidance, Identification, Gauging, and Inspection

Machine vision increases quality and productivity in reduced costs. If correct camera resolution and optics are selected, then machine vision system can easily identify and analyze very small details as well which otherwise was not recognized by human eyes. Apart from this, machine vision brings safety benefits also by

Machine Vision Technology, Deep Learning, and IoT in Agricultural Industry 147

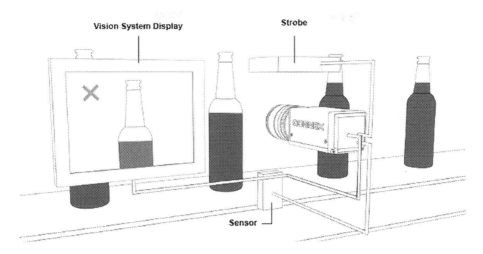

FIGURE 8.2 Bottle fill-level inspection. *Source: www.cognex.com.*

reducing involvement of human beings in manufacturing environment that sometimes is hazardous for labors

8.2.1 COMPONENTS OF MACHINE VISION

The major components of a machine vision system include the lighting, lens, image sensor, vision processing, and communication system. Lighting system is required to provide light on the part to be inspected. Camera can easily capture images from brightened parts with good quality lenses which forward them to sensors for being converted into digital images. Images are passed on to vision processing system for analysis and extraction of features. On the basis of extracted features, necessary decisions are made and communicated to devices. Functionality is shown in Figure 8.3.

Most of the machine vision hardware components are available commercially. Components can be assembled to make single device. Factors affecting the performance of the system in image capturing are sensitivity and resolution. Both are interdependent. Increasing sensitivity reduces resolution and vice versa. Sensitivity is capability of vision machine to work in dim light. Resolution helps machine in differentiating between two objects. Better the resolution, better is the vision. Choosing the right vision system to meet the requirement of the application is very crucial.

FIGURE 8.3 Components of machine vision.

8.2.2 Machine Vision in Agriculture

Agriculture sector is also making use of machine vision technology in land- or areal-based remote sensing of assessment of natural resources, crop species and quality detection, precision farming, etc. AI programming can also be incorporated with machine vision system for better performance. Latest tools like neural networks, expert systems, and fuzzy logic can also be applied.

8.3 DEEP LEARNING AND ITS TECHNIQUES

DL mimics the working of human brain for data processing and pattern identification for decision-making. DL technology has emerged as prominent technology in medical diagnosis, text, speech and facial recognition, Internet security, mobile and wearable devices, agriculture, and manufacturing applications where accurate decision-making is significant.

Some popular DL algorithms are feed forward neural network with back propagation (BP), convolution neural network (CNN), recurrent neural network (RNN), and generative adversarial network (GAN). Applications of these DL algorithms are widely apparent in the field of agriculture.

8.3.1 Feed-Forward Neural Network and Back Propagation (BP)

In the field of Artificial Neural Networks Feed-Forward Back Propagation algorithm is a supervised learning way of multilayer networks. Entire model is created from different layers with clear input and output layers and multiple hidden layers as shown in Figure 8.4. Internal weighs of input signals are adjusted multiple times by according to the errors in produced output and expected output [6]. Errors are minimized to acceptable level.

8.3.2 Convolutional Neural Networks (CNN/ConvNet)

CNN model are inspired by connectivity pattern of neurons in human brains [7]. Convolution networks are form of DL algorithms that are designed for image inputs. Networks is designed to provide learnable weights to various detectable objects in picture input. Image is flattened to make an array and considering value of pixel as features to get the number in image. This is called weight matrix and it behaves like a filter. Weight might be extracting edges or a particular color. A convolutional network consists of convolutional layer, optional pooling layer and output layer. If image is very big, number of trainable parameters are reduced. That's why pooling layer is included between two convolutional layers to decrease the spatial size of the image as shown in the Figure 8.5.

8.3.3 Recurrent Neural Networks (RNN)

As name suggests, recurrent neural networks feeds the output from last executed step in current step [8]. In conventional neural networks all input and output are independent of one another. But many times it is required to remember the previous

Machine Vision Technology, Deep Learning, and IoT in Agricultural Industry 149

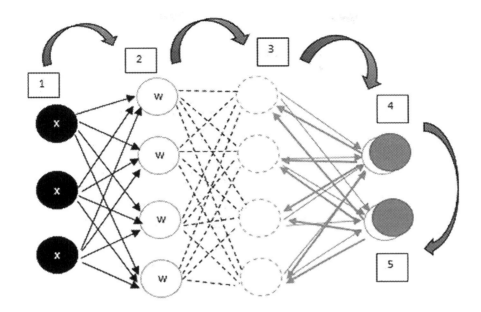

FIGURE 8.4 Simple feed-forward network with back-propagation.

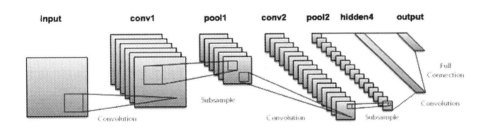

FIGURE 8.5 Convolutional neural network.

state. For example to predict the next word of the sentence, words written before are needed in order to give suggestions. This requirement of memory is fulfilled with the help of hidden layer in RNN. It is only this hidden layer used for remembering the state makes RNN different than other algorithms.

8.3.4 GENERATIVE ADVERSARIAL NETWORKS (GAN)

Under the category of unsupervised learning another powerful class of neural networks is Generative Adversarial Networks. It makes use of probabilistic models to generate data and thus is Generative. It is Adversarial as training in done under adversarial settings. And for training purpose it makes use of deep neural network algorithms [9]. GAN consists of generator and discriminator. Generator produces mock samples of image or audio data and aims to dodge discriminator.

150   Industrial Internet of Things

While discriminator is responsible for differentiating real and fake samples; both, Discriminator and Generator are designed as neural networks and proceed in contest with each other during their training stage. The whole process is iterated numerous times and performance of discriminator and generator keeps on improving after each pass in loop.

8.4 ADVANTAGES OF COMBINING MACHINE VISION AND DEEP LEARNING

Machine vision systems are reliable for well manufactured parts or well identifiable features. But with the unexpected exceptions or defects in the system, it becomes difficult to program. Some conventional machine vision inspections are difficult to perform because of certain variables which are hard to be identified such as change in color, lighting, curvature, etc. In such cases, DL-based software offers better alternative as they are both robust as well as flexible. Generally the choice between machine vision and DL depends upon the type of application, volume of involved data, and processing capabilities. Rule based techniques are useful for gauging, measuring and alignment kind of things while DL is used for classification of material with inherent defects also.

Sometimes application may need involvement of both the technologies. For example, in fixture of region of interest ML may be applied and for inspecting the region DL algorithm may be used. Result of later is passed backed to vision system to take accurate measurements of the defect size and shape. Traditional computer vision when combined with DL creates better AI. For Internet of Things (IoT) applications, machine vision forms basis. For example, house holding cameras are used to inspect activities going on at home in the absence of family members. With the improvements in device capabilities in terms of capacity, power and optics, performance of IoT applications has also been increased and become more cost-effective. Recently, advancements in AI and specifically in DL have accelerate the proliferation of vision based applications in IoT with great accuracy in task of classification. DL makes use of neural networks which are trained not programmed so the applications using this approach are easier to be developed and take more advantage of enormous amount of data available in the forms of images or videos. DL has better versatility as well because frame works can be reused as compared to machine vision algorithms which are not general purpose.

Mixing machine vision with DL is giving better performance by combining advantages of both the techniques without trade-offs.

8.5 IoT SOLUTIONS TO AGRICULTURAL PROBLEMS

IoT that is Internet of Things can add value to all areas of farming to make it smart farming exactly in the same way in which other connected devices have entered every aspect of our daily life. With the use of IoT-enabled gadgets, agriculture has been industrialized and has become more technology driven as farmers have got better control over livestock and crops. Now when driverless cars and virtual reality are no more science fiction, all these are used in everyday occurrences. According to

Machine Vision Technology, Deep Learning, and IoT in Agricultural Industry   **151**

recent statistics in 2020, the market share for IoT in agriculture reached $5.6 billion. At the same time, the global smart agriculture market size is expected to triple by 2025, reaching $15.3 billion. Application of IoT solutions in agriculture has led to smart farming. IoT sensors are used to collect different environmental and machine metrics which are used by farmers to make decisions more precisely and to improve every aspect of their work. Sensors can help farmers to monitor the state of the crop so that he can decide upon the quantity and quality of fertilizers and pesticides. IoT is transforming the future of agriculture. IoT solutions are focused on helping farmers close the supply demand gap, by ensuring high yields, profitability, and protection of the environment. Farmers have started to realize that the IoT is a driving force for increasing agricultural production in a cost-effective way.

8.5.1   ADVANTAGES OF IOT IN SMART FARMING

Tremendous agricultural transformations can be made with the use of technology and IoT in fields. IoT in the field of agriculture is considered as second green revolution and its advantages to farmers are two folds. Farmers can decrease their cost involved in process of farming and increase the yields at the same time by making precise decision on accurate data provided.

- Smart agricultural sensors can be used to collect huge amount of data related to weather conditions, soil quality, growth of crop, plant diseases, health conditions of cattle, etc., which can later be used to track the state of agricultural productivity.
- Use IoT may help in lowering down the production risks. Production can easily be predicted which serve as the basis for further planning as now one is aware of crop quantity to be harvested
- With automated processes business efficiency increases. With the use of smart devices, one can automate multiple processes across the complete production cycle, e.g., irrigation, fertilizing, or pest control.

Consequently higher revenues can be generated.

8.5.2   IOT USE CASES IN SMART FARMING

IoT sensors in agriculture are of many types and responsible for collecting data about various crop conditions. Monitoring of equipment, cattle, crops, etc., can easily be done by farmers with the help of their smart Farmers use this technology to run statistical predictions for their crops and livestock.

8.5.2.1   Climate Conditions Monitoring

Role of climate in farming is very crucial and it directly affects the quality and quantity of crops. IOT can provide farmers with the real time weather conditions. Number of sensors are placed at different locations inside and outside of fields responsible for collecting data from environment like temperature, rainfall, humidity, etc. This information can be used to select crop which can sustain under particular climate conditions. For changing and disturbing weather conditions SMS alerts are sent. Few such kind of agricultural IoT devices are allMETEO, Smart Elements, and Pycno.

8.5.2.2 Automation of Greenhouse

Real-time greenhouse conditions like temperature, humidity, soil, and light can be collected using IoT sensors, thus eliminating the manual intervention of farmers thus making entire process cost-effective and increasing accuracy at the same time. Conditions can also be adjusted to match the set parameters. Farmapp and Growlink are IoT agriculture products offering such capabilities. GreenIQ is also an interesting product that uses smart agriculture sensors. It is a smart sprinklers controller that allows you to manage your irrigation and lighting systems remotely.

8.5.2.3 Management of Crops

Arable and Semios can serve as good representations of how this use case can be applied in real life. Manufacturers have started connecting tractors to the Internet and have created a method to display data about farmers' crop yields. Self-driving tractors would free up farmers to perform other tasks and further increase efficiency.

8.5.2.4 Cattle Monitoring and Management

IoT agriculture sensors can be attached to cattle to monitor their health and performance. Tracking and monitoring of livestock help in gathering data related to health, well-being, and location of the livestock. If any sick animal is identified, then it can be separated from herd to avoid spread of disease. Drones can be utilized to track the physical location of animals. SCR by Allflex and Cowlar use smart agriculture sensors (collar tags) to deliver temperature, health, activity, and nutrition insights on each individual cow as well as collective information about the herd. One of the solutions helps the cattle owners observe cows that are pregnant and about to give birth. From the heifer, a sensor powered by a battery is expelled when its water breaks. This sends information to the herd manager or the rancher. In the time that is spent with heifers that are giving birth, the sensor enables farmers to be more focused.

8.5.2.5 Precision Farming

The approach of using IoT technology to ensure optimum application of resources to achieve high crop yields and reduce operational costs is called precision agriculture. Precision farming refers to efficiency in making accurate decisions with the help of collected data. IoT sensors can collect a wide variety of metrics related to different field conditions like temperature, humidity, light conditions, CO2 levels and pest infections. With the availability of this data farmers can decide the optimal needs of crops in terms of fertilizers, pesticides, water etc. By precisely measuring variations within a field, farmers can boost the effectiveness of pesticides and fertilizers, or use them selectively. This leads to better and healthier crops at reduced costs. For example, CropX builds IoT soil sensors that measure soil moisture, temperature, and electric conductivity enabling farmers to approach each crop's unique needs individually. Mothive offers similar services, helping farmers reduce waste, improve yields, and increase farm sustainability. Similar solutions are represented by FarmLogs and Cropio

8.5.2.6 Agricultural Drones

Drones are also known as UAVs that is unmanned aerial vehicles. Drones can perform a wide range of tasks where human intervention was required earlier like planting plants, detecting pests and infections, medicine spraying, crop monitoring, etc., apart from surveillance. DroneSeed, for example, builds drones for planting trees in deforested areas. The use of such drones is many folds more effective than human labor. A Sense Fly agriculture drone eBee SQ uses multispectral image analyses to estimate the health of crops and comes at an affordable price. Drones while flying can collect multispectral, thermal and visual images and data collected can give farmers complete insight of plant health, fruit counting, yield prediction, plant height, chlorophyll measurement, contents in crops, drainage mapping, weed growth, etc.

8.5.2.7 Predictive Analytics for Smart Farming

Predictive data analysis is first step for precision agriculture. The use of data analytics in real time data collected through different sensors helps farmers in anticipating some important factors like time of harvesting, risk of diseases and infections, volume of yields, etc. Now agriculture management has become easier with available data analytics tools. If farmer comes to know about unexpected climate conditions in advance like floods or droughts then accordingly he can optimize the supply of nutrients and water to crops to improve quality. SoilScout is one of such solutions available, which helps farmers in saving 50% of irrigation water and delivering actionable insight regardless of climate conditions and season.

8.5.3 Challenges in IoT Based Smart Farming

Although there are many use cases for IoT in agriculture, smart devices and sensors are performing well to provide farmers with real time data, which are assisting farmers to make accurate predictions and hence increasing performance and revenue. But lots of challenges are also existing that needs to be considered before going for smart farming.

8.5.3.1 Sensor Selection

IoT solutions in farming are successful if correct sensors are selected for devices. Sometimes according to the information to be gathered they are to be designed as well. Quality of sensors is also crucial as performance of the IoT solution depends largely on the accuracy and reliability of the collected data

8.5.3.2 Appropriate Data Analytical Tools

Apart from reliable data collection, accurate and powerful data analytics is also required for IOT solutions. Strong data analytics, application of predictive algorithm, and ML algorithms are required to obtain actionable insights on collected data otherwise data is of no use if it cannot be appropriately sensed.

8.5.3.3 Maintenance of IoT Setup

The most challenging part of IoT solutions in agriculture is the maintenance of hardware setup and specifically sensors deployed in fields. They are more prone to

154 Industrial Internet of Things

damage and can be broken easily. Selected hardware should be durable and easy to maintain.

8.5.3.4 Connectivity

Data transmission is the backbone of entire IoT setup for the adoption of smart farming. Connection between agricultural facilities should be reliable enough to withstand the changing and sometimes poor weather conditions so that every operation goes without any disruption. Upcoming internet technologies may help in this aspect. Without appropriate internet connectivity capable of working in changing weather conditions IoT setup is not going to perform as required and expected.

8.5.3.5 Data Security

Huge amount of data is involved in application of precision agriculture and IoT technology which further increases the security risks. Probability of data theft and hacking attacks increases with security loops in data management. Not much work has been done in this field yet. Precision agriculture and IoT technology imply working with large sets of data, which increases the number of potential security loopholes that perpetrators can use for data theft and hacking attacks. Unfortunately, data security in agriculture is still, to a large extent, an unfamiliar concept. Different forms of data transmissions are involved, different machines, sensors, drones, robots are connected to Internet but with almost zero protection. Sometimes not even authentication process for remote access by farmers. To ensure integrity of basic IoT application encryption methods may be applied to protect sensitive data traffic monitored. AI-based security tools may be used to detect suspicious activities in real time or storing data in block chain may also be used.

8.5.3.6 Data Collection Frequency

Wide variety of data types are available in agriculture industry and ensuring perfect data collection frequency is difficult. If collected data from fields are not delivered and shared timely, they become useless and destroy the purpose of entire setup.

Thus, the IoT agricultural applications are making it possible for farmers to collect meaningful data, analyze it and take action accordingly.

8.6 AGROBOTS – AGRICULTURAL ROBOTS

Designing of agricultural robots that are agrobots are highly dependent on modern technologies like sensor technology; sensors are the major tool that is required in collecting the real time data and using these data we train the robots to perform the activities we want it to do. Most of the agrobots are based on the image processing techniques these agrobots may include the fruit picking robots which process the image of a fruit and then decide if it is ripen or not. After processing that the fruit is ripened scans the path of the robotic arm to pluck the fruit. Vision-based processing also plays a major role in in the process of seedling, weeding and target spraying.

Use of agriculture robots in fields has been proposed by many researchers. In 2018, Bosilj et al. described a methodology of classification of different types of crops by

Machine Vision Technology, Deep Learning, and IoT in Agricultural Industry 155

their image processing [10]. Different types of morphological characteristics are collected with high-definition cameras and based on the images collected the crops are classified to their respective category and proposed the soil required for it, different types of weeds that can infect the plants. Ramin Shamshiri et al. in their work [11] mentioned the use of IoT in collection of data about crop, then analyzed and visualized the data to predict the type of resources required and appropriate time that will help with the better planning of steps of cultivation by analyzing the current climatic conditions. Time to time different surveys on the use of robots working independently in fields have been proposed and analyzed [11–16].

8.6.1 Agrobots Working in Fields for Different Agricultural Activities

Weed control and targeted spraying robots: The main reason to build agrobots is to reduce the involvement of humans in agriculture field (Figure 8.6(a–d)). Introduction of robots for weed control is both effective for time and labor.

(a) BoniRob is a robot that can create a detailed map of field and is effective for spraying weedicides for the row crops.
(b) Agbot is another robot which is made to detect the weeds and is emancipated in application of fertilizers. It has algorithms for classification of different weeds and then manually or chemically removing it.
(c) Tertill is an agricultural robot that uses solar energy for its operation and is made for weed cutting.

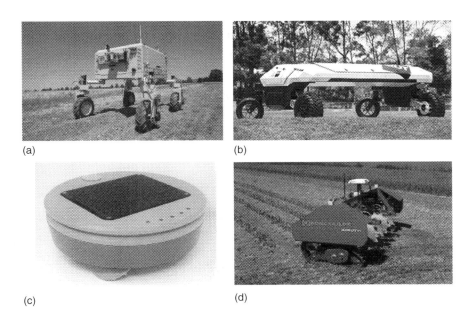

FIGURE 8.6 (a) Bonirob, source: *www.agriexpo.online* (b) Agbot, source: *www.qut.edu.au* (c) Tertill, source: *www.franklinrobotics.com* (d) Rootti, source: *www.kongskilde.com*.

(d) Kongskilde Rootti is a based on FoboMind and is capable of doing fully or semi-autonomous precision seedling system, weed control and also has furrow opening and cleaning.

Field scouting and data collection robots. These robots have sensors attached with them to collection of data which is very tedious task and has a lot of reliability issues, see Figure 8.7(a–d). The devices used for data collection must provide veracious information. If these robots are accurately assembled, then these will prove its worth in terms of flexibility and multipurpose use and thus will proved to be very cost effective. The devices needed to be equipped with advanced imaging sensors that work with 3D point clouds and also with GPS navigation system.

(a) Trimbot is an autonomous navigation and monitoring robot which is specifically used in rose plantation area. Being equipped with robotic arm these robots are designed for rose cutting and trimming of rose bushes.
(b) Wall-Ye is an agrobot specially built for vineyards. It is designed for mapping the grapes through camera and sensors, pruning and harvesting the grapes.
(c) Ladybird is used for multiple purposes and is also an autonomous robot used for monitoring, have sensors or mapping objects detected during surveillance and uses advanced classification algorithm for distinguishing between different varieties of vegetables.

FIGURE 8.7 (a) Trimbot, *source: www.sciencephoto.com* (b) Wall-yee, *source: wall-yee.com* (c) Ladybird, source: robohub.org (d) MARS, *source: fendt.com.*

Machine Vision Technology, Deep Learning, and IoT in Agricultural Industry **157**

(d) MARS (Mobile Agricultural Robot Swarm) consists of swarm of few small and sleek robots. The main idea of designing such robots is to minimize soil compaction and also to reduce energy consumption. These robots can be customized for farm specific use.

8.6.2 HARVESTING ROBOTS

Harvesting is an intensive, time-consuming and expensive process of farming. Harvesting is one of the most censorious phases of agriculture, and it poses different challenges. Besides that agricultural harvesting process largely depends on labor availability. So there's a big demand at the process to move to using some autonomous machines. Few of the robots harvested are Harvey, CROPS, SWEEPER, Energid citrus picking system etc. Most of these robots are emanate on the principle of detect and move. These robots are equipped with high quality of sensors, camera and a robotic arm. The robots detect fruits through camera and sensors. The next step is to move the robotic arm to the fruit so as to pluck it. The algorithms are designed so scan the shortest path between the fruit and robotic arm considering the fact that the branches should be always at the back of the fruit.

8.6.3 CHALLENGES IN IMPLEMENTING AGROBOTS

After several years of research now scientists have been able to develop advanced and efficient sensors for collecting information. But data collection is not always ample, it only helps in making better analysis and decision-making, which only reduces our decision-making time but can only provide profit if it is supported by efficient management technologies. One another challenge of using networking in this process is the threat of network being compromised. If the network is being compromised it will cause catastrophic loss for farmers as will create ambiguity in data collected by the farmer. When algorithms are applied on these ambiguous data will obviously produce wrong decisions and predictions and thus will cause a great loss to the farmer.

The next problem in implementing is huge investment. These technologies are very beneficial but often benefits come with a cost. Installing robotics equipment in fields comes with a heavy on time investments because the designing of such efficient robots requires an obligatory skill and expertise and also the software used in these robots are very expensive. Sometimes the system may confuse about the output and may give a wrong output. One of such example is analysis of color of leaf and detecting the disease in plant. The color of leaf may turn yellow due to normal leaf shedding process by a healthy plant or due to lack of nitrogen content in soil or because of lack of moisture. This situation requires machine to interact with human to produce a correct output. But interaction with the humans is not desirable for developing fully automated system. The other problem with its implementation in developing countries like India is connectivity as the agricultural fields are located in the village areas of the country. The village areas have very poor wireless connection. The main problem with the wireless connectivity signals. We cannot think of robotics system like MARS to be implemented in such interior village areas of a country. The other challenge is withstanding the wrath of nature some robots might work appropriately in test conditions in laboratories but the things may not work when it comes to a real-time situation. Some of the sensors may even get damaged due to heat or not work appropriately work in excessive cold weather. Withstanding in rain is another a tough task. Robots designed need to be waterproof and be light so that it can be taken out of the field during rain from the marshy soil or need to be implemented with the wheels that could drive robots out of the marshy area. Maintenance is also a big challenge for the robots because in country like India most of the farmers lack technical knowledge required for the maintenance of the robots. They need either to depend on authorized company for its maintenance or need to hire some technical expert for its maintenance. This will again add up to the cost of total agricultural production. There has been oddles of research taking place about developing advanced agricultural robots in past few years. The introduction of robots in agriculture has greatly reduced the dependence of framers on labors and helped to overcome the labor shortage problem. With more advanced tools it provides correct decision steps like accurate detection of diseases or find the requirement of fertilizers in plant, detection and removal of weeds. Robots also involve human interaction to come-up with a fully correct outcome.

8.6.4 CONCLUSION AND FUTURE WORK

Smart farming is the need of hour. To improve production in terms of both quality and quantity, application of AI in different fields of agriculture is clearly evident and cannot be ruled out. And when supported with machine vision technologies and IOT good results can be achieved. Farmers would be highly benefitted from advanced technologies. This chapter explained each and every aspect of application of AI in smart farming. DL and CNN proved out to be the best architecture for image processing which is an integral part in application of AI in agriculture. Still researches are going on to optimize the DL models to achieve better results and wonders may be expected in near future.

REFERENCES

[1] D.N. Baker, J.R. Lambert, J.M. McKinion, "GOSSYM: a simulator of cotton crop growth and yield." *Technical Bulletin, Agricultural Experiment Station*, South Carolina, USA, NCpedia, 1983.

[2] P. Martiniello, "Development of a database computer management system for retrieval on varietal field evaluation and plant breeding information in agriculture." *Computers and Electronics in Agriculture*, vol. 2, no. 3, pp. 183–192, 1988.

[3] K. W. Thorpe, R. L. Ridgway, R. E. Webb, "A computerized data management and decision support system for gypsy moth management in suburban parks." *Computers and Electronics in Agriculture*, vol. 6, no. 4, pp. 333–345, 1992.

[4] P. J. Ramos, F. A. Prieto, E. C. Montoya, C. E. Oliveros, "Automatic fruit count on coffee branches using computer vision." *Computers and Electronics in Agriculture*, vol. 137, pp. 9–22, 2017.

[5] J. M. McKinion, H. E. Lemmon, "Expert systems for agriculture." *Computers and Electronics in Agriculture*, vol. 1, no. 1, pp. 31–40, 1985.

[6] A.T.C. Goh, "Back-propagation neural networks for modeling complex systems." *Artificial Intelligence in Engineering*, vol. 9, no. 3, pp. 143–151, 1995.

Machine Vision Technology, Deep Learning, and IoT in Agricultural Industry **159**

[7] A.S. Razavian, H. Azizpour, J. Sullivan, S. Carlsson, *"CNN features off-the-shelf: an astounding baseline for recognition."* In: *IEEE Conference on Computer Vision and Pattern Recognition Workshops (CVPRW)*, 2014.

[8] Y. LeCun, Y. Bengio, G. Hinton, "Deep learning." *Nature*, vol. 521, no. 7553, p. 436, 2015.

[9] A. Radford, L. Metz, S. Chintala, *"Unsupervised representation learning with deep convolutional generative adversarial networks."* In: *ICLR*, San Juan, Puerto Rico, 2016.

[10] P. Bosilj, T. Duckett, G. Cielniak, "Analysis of morphology-based features for classification of crop and weeds in precision agriculture." *IEEE Robotics and Automation Letters*, vol. 3, no. 4, pp. 2950–2956, 2018.

[11] R. Ramin Shamshiri, C. Weltzien, I. A. Hameed, J. Yule I, E. Grift T, K. Balasundram S, G. Chowdhary. "Research and development in agricultural robotics: a perspective of digital farming." *International Journal of Agricultural and Biological Engineering*, vol. 11, no. 4, pp. 1–11.

[12] M. Behmanesh, T. S. Hong, M. S. M. Kassim, A. Azim, & Z. Dashtizadeh, "A brief survey on agricultural robots." *International Journal of Mechanical Engineering and Robotics Research*, vol. 6, no. 3, pp. 178–182, 2017.

[13] K. Jensen, M. Larsen, S. H. Nielsen, L. B. Larsen, K. S. Olsen, and R. N. Jørgensen, "Towards an open software platform for field robots in precision agriculture." *Robotics*, vol. 3, no. 2, pp. 207–234, 2014.

[14] UK RAS Network. The Future of Robotic Agriculture. UK-RAS White Papers. Retrieved from www.ukras.org, 2018.

[15] M. R. Benjamin, J. J. Leonard, H. Schmidt, and P. M. Newman, "Nested autonomy for unmanned marine vehicles with MOOS-IvP." *Journal of Field Robotics*, vol. 29, no. 4, pp. 554–575, 2012, doi:10.1002/rob

[16] A. Amrita Sneha, E. Abirami, A. Ankita, R. Praveena, and R. Srimeena, *"Agricultural robot for automatic ploughing and seeding."* In: *Proceedings - 2015 IEEE International Conference on Technological Innovations in ICT for Agriculture and Rural Development, TIAR 2015, (Tiar)*, Chennai, India, pp. 17–23, 2015.

9 IIoT Edge Network and Spectrum Scarcity Issue

Gyanendra Prasad Joshi
Sejong University, Seoul, South Korea

Sudan Jha
CHRIST University, Delhi-NCR Campus, Delhi, India

CONTENTS

9.1 Introduction 161
    9.1.1 IoT and IIoT 162
9.2 IIoT Edge and Edge Devices 163
9.3 Computing Strategies for IIoT 165
    9.3.1 Cloud Computing 165
9.4 Edge Computing 166
9.5 Fog Computing and Hybrid Techniques 167
9.6 Connectivity on IIoT 167
9.7 CR for Future IIoT 168
    9.7.1 Spectrum Scarcity Problem in IIoT and Cognitive-IIoT 168
    9.7.2 Cognitive LPWAN for IIoT 171
9.8 Conclusion 172
References 173

9.1 INTRODUCTION

The industrial Internet of Things (IIoT), smart factory, and smart manufacturing are new exciting terms for this generation. These are the application of the Internet of Things (IoT) that supports the realization of Industry 4.0. Because of advancements in wireless networks and sensors, millions of industrial devices are connected, gather, and share data in asset-intensive industries such as manufacturing, telecommunications, mining, construction, waste and water management, and energy generation and distribution. Because advancement in microchip technologies, even tiny and cheap devices give a high level of digital intelligence that collects detailed data, processes it, and sends it, that ultimately allows machinery, manufacturing processes, services, and people to be tracked and monitored. Data generated by these devices can be used for business processes, which helps to boost efficiency, productivity, and safety in smart manufacturing and distributing process.

DOI: 10.1201/9781003102267-9

With the growth and adoption of IoT, factories are becoming more instrumented and interconnected. The extremely detailed data collected from IIoT devices are very important for faster and better decision-making. These data help companies understand their business process and give an insight into the border supply chain. The IIoT brings realization of digital transformation by digital convergence in business. Recently, the IIoT is of particular interest to the smart manufacturing, and factory automation of oil and gas industries, agriculture industries, chemical plants, mines, and transportation and logistics industries. In fact, the IIoT is not only a single technology but also a combination of different techniques, such as IoT, big data, cyber-physical systems (CPS), machine learning (ML), and simulation, to organize smart operations in the industrial environments (O'Donovan et al., 2015).

9.1.1 IoT AND IIoT

Although there is some overlap between IIoT and IoT, the consumer IoT, or just called IoT, is the network of physical objects that are equipped with sensing or tracking technologies, and software to connect and share data with other devices and systems. Some examples of IoT devices are smartwatches, smart home speakers, light bulbs, door locks, smart white goods, and smart brown goods. The IIoT refers to interconnected sensors, machines, tools, instruments, process, and other industrial devices networked together with industrial applications, with the major objective to achieve high operational efficiency, increased productivity, and better management of industrial assets and processes through product customization, intelligent monitoring applications for production floor shops and machine health, and predictive and preventive maintenance of industrial equipment (Khan et al., 2020). The IIoT connects people, products, and processes to power digital transformation in industries. The basic objective of both IoT and IIoT is the same, i.e., to use sensors and automation to make processes more efficient. However, the consumer IoT is about making life easier and IIoT is for an intelligent and efficient industrial process. Boyes et al. discussed the definitions of IoT and IIoT and gaps in the current literature and understanding of IIoT (Boyes et al., 2018).

Although tracking and monitoring goods using sensors and RFIDs are not a new thing, the utilization of these is emerged with the evolution of IIoT, the low price of sensors, pervasive wireless networking options, and the advent of big-data analytics (Ranger, 2019). These devices are sophisticated IoT devices that perform some degree of data processing within the device itself. Recently, use of an intelligent edge device is significantly increasing in industries, and many industrialized countries such as the US, China, Japan, Germany, Korea, France, and the UK are spending billions of dollars for IIoT (Snyder et al., 2020; Espinoza et al., 2020).

Although it is already discussed in other chapters, the IIoT is often confused with and interchangeably used for machine-to-machine (M2M), "fourth industrial revolution", or Industry 4.0, among others. Basically, M2M or M2M communication is a specialized form of data communication that allows machines to autonomously communicate with each other and exchange information required for business operations. Basically, M2M provides a platform to run the IIoT. Industry 4.0 or the fourth industrial revolution refers to the automation of conventional production and industrial practices using current smart technologies. A large-scale AI, analytics, M2M communication, and IIoT are incorporated in Industry 4.0 for efficient automation and connectivity for diagnose problems and enhance the business process.

Industries often look to the IIoT to bridge the gap between information technology and operational technology (Paine, 2017). The main components of IIoT are the sensing and tracking devices, networks, middleware and applications, and analytics. Figure 9.1 shows the IIoT architecture including IIoT edge devices, tools, platforms, and applications in layers. In the following sections, IIoT edge networks and spectrum scarcity issues in IIoT are discussed.

9.2 IIoT EDGE AND EDGE DEVICES

The IIoT is composed of interconnected machines and devices that can monitor, collect, exchange, and analyze data. The edge devices are the devices that work at the edges of the networks. Edge devices encompass a broad range of device types, including sensing, tracking, monitoring, processing, transmission, routing, storage, routing, integrated accessing, multiplexing, and edge gateway devices. These network edges can host thousands of machines that contain crucial industrial data.

The sensing devices can be anything from photosensors (i.e., cameras) to vibration, humidity, heat sensors (i.e., thermometers), or motion sensors, depending on the industry. Various types of tracking devices are used in industries, such as radio frequency identification devices (RFID), real-time locating systems (RTLS), and global positioning systems (GPS). The RFID uses small tags containing a microchip or transistor with encoded information and an antenna for receiving and sending signals. Generally, passive tags are used; however, based on the industry active tags can be used by adding a power source to an RFID tag. These "active" trackers can respond to a wide-band scanning signal. It can also set up to go off periodically for energy conservation. The GPS trackers are rarely used in IIoT, because of their capacity to tracking indoor. Several other indoor localization systems are used for indoor localization, such as Proximity-based Systems, Wi-Fi-based Systems, Ultra-Wide-Band (UWB) Systems, Acoustic Systems, and infrared (IR)-based systems, among others.

One of the many capacities of the IIoT is that it helps companies generate massive amounts of data. However, these data are only valuable if it can be accessed and acted upon swiftly, efficiently, and securely. Although wired networks connect many devices in the industries, several mobile devices may require efficient manufacturing process management (MPM). The network used for IIoT can range from Bluetooth and Wi-Fi though to 5G networks and satellite depending on the industry type. Because the data generated by the devices for operation are crucial and confidential many times, these wireless devices also require security functions. Edge devices, such as wireless access points (APs) or virtual private network (VPN) servers, should have security capabilities to block malicious users or device connections.

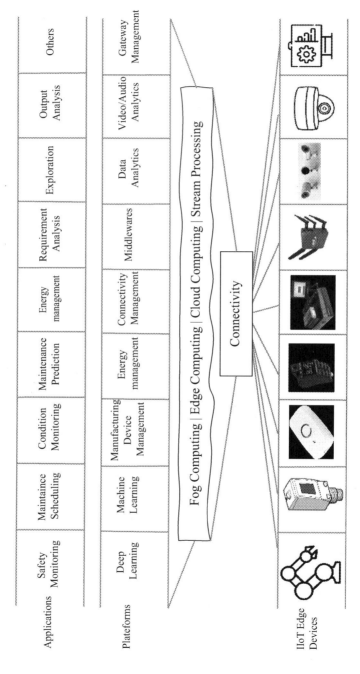

FIGURE 9.1 IIoT architecture including IIoT edge devices, tools, platforms, and applications in layers.

9.3 COMPUTING STRATEGIES FOR IIoT

Data generated by IIoT devices can be processed in the industry premises; however, these IIoT devices may have limited computing capacities, thus the need to be processed in the more powerful systems in the remote cloud. Basically, IIoT is carried out in the following types of computing strategies: cloud computing, edge computing, and fog computing.

9.3.1 Cloud Computing

Cloud computing is the on-demand computing environment that can dynamically scale to meet a set of changing demands. All data are collected and stored at a centralized server in a data center. Any devices that need to access the data or related applications in the cloud server need to connect to the cloud server via communication networks. Because of the centralized control cloud is generally easy to secure and control.

Industrial cloud computing is a digital and Internet-based process that connects systems, people, and industries. Industrial cloud computing provides the infrastructure for the data transmitted to the cloud servers. It also provides software applications for the devices in the industry to automate workflows. With industrial cloud computing, organizations are able to organize and analyze massive data generated in the industry and apply them more strategically in the overall business operations.

A cloud solution serves an application by providing everything as a service, such as SaaS (Software as a Service), IaaS (Infrastructure as a service), and PaaS (Platform as a service) (Farnell, 2016).

a. Infrastructure as a Service (IaaS): IaaS is a cloud computing model that provides virtualized computing resources over the Internet. The IaaS is used to deploy PaaS, SaaS, and web applications. End users do not have control over the IaaS, but they are able to access the servers and storage and can operate a virtual data center in the cloud. There are so many IaaS providers around the globe. Some examples of IaaS providers are Amazon Web Services (AWS), Windows Azure, Google Compute Engine, Rackspace Open Cloud, and IBM SmartCloud Enterprise. Currently, GE's Predix platform for the IIoT is available to run on Microsoft Windows Azure Cloud (Microsoft News Center, 2016).

b. Platform as a Service (PaaS): PaaS is a cloud computing model in which the vendor provides a software platform to the developers for developing software. With PaaS, developers can develop, test, and deploy applications easily and efficiently. End users cannot control or manage the underlying infrastructure, but they can control the applications and configuration of the application-hosting environment. GE's Predix, Honeywell's Sentience, and Siemens's MindSphere are examples of PaaS's for industrial applications (Rahman et al., 2019).

c. Software as a Service (SaaS): SaaS is a form of cloud computing that delivers a cloud application—and all its underlying IT infrastructure and platforms—to users via a web browser or a program interface. The advantage of a SaaS is that end-user need to run or install specific applications on individual computers. Also, end-users do not need to maintain infrastructure, platforms, and software. Few examples of SaaS are Google Apps, Cisco WebEx, and Siemens's Industrial Machinery Catalyst on the Cloud.

(Siemens PLM Software, 2021)

Since cloud computing is built on a scalable infrastructure of data centers, it can extend its storage and processing capacity if needed. This scalability is a huge advantage for small companies looking to grow rapidly. The centralized structure of cloud computing makes it difficult to process data collected from the IIoT edges quickly and efficiently. Although it has advantages of scalability on computing power and storage capacity, the cloud lacks speed.

9.4 EDGE COMPUTING

Fundamental issues associated with the centralization of cloud architecture are addressed by IIoT edge. While clouds are powerful for storage and processing, as discussed earlier, they cause delays for IIoT devices sending data back and forth. Data from these remote locations have the potential to generate valuable business, but for time-critical operations, it is often too far away, too costly, or too unreliable to transmit.

Edge and fog computing are two potential solutions for the massive influx of data collected from IIoT devices. Edge computing keeps computing capabilities to local devices. Data can be processed quicker by IIoT edge devices, avoiding delays, security breaches, and other issues. The difficulty of providing data in realtime is addressed by IIoT edges.

For example, if a machine is just started malfunctioning and needs to stop the machine. If the machine must send data to the cloud and receive instructions to stop the machine, it may lead to defective products or a big accident. In this type of scenario, instead of sending such an important and emergency data to the cloud and waiting for the action to take, an IIoT edge device can give quick decisions locally by eliminating the roundtrip time required for cloud processing.

Edge gateways manage edge computing and data storage. It hosts localized and task-specific activity in order to analyze edge data in near real-time. Therefore, very little data are transferred to the main server, which saves the bandwidth requirements. Moreover, these gateways mitigate security concerns by retaining confidential and sensitive data within a local network and analyzing it within a secure system. Many edge computing devices, such as MQTT, OPC UA, AMQP, and CoAP for secure communications, use advanced communication techniques (Tightiz & Yang, 2020).

One of the key advantages of implementing edge computing is the ability to capture and analyze data as it is collected, analyzing and addressing issues that could not be detected as easily if the data were to be sent for processing and analysis to a centralized cloud server. The security risk associated with data porting is minimized by keeping data on the edge without sending it to the remote server. By processing some data on-site also reduces bandwidth costs. If an industry has hundreds of IIoT devices, it is not efficient to attempt to connect to the remote resource at once. Edge devices capture, process, and store data closer to endpoints in a more distributed manner. It also accelerates response time, minimizes latency, and conserves network resources.

Cloud computing can be a right choice when an industry requires huge storage and high processing power to run certain applications and processes. On the other hand, when latency is an issue, and local actions are required, edge computing is a better option. Also, when backend traffic needs to reduce and confidential data is involved edge computing can be a correct option.

9.5 FOG COMPUTING AND HYBRID TECHNIQUES

Fog computing is an alternative approach to cloud computing models to deliver services efficiently and quickly. In fog computing, data are processed within a fog node or an IIoT gateway, which is situated within the local area network. However, the data are processed on the IIoT devices without sending the data to remote devices in the cloud as in cloud computing. If any industry has millions of connected devices sharing data back and forth, the fog computing approach can aggregate data, and process it with a larger capacity.

Depending on the nature of the industry, in IIoT edge, fog and cloud computing can be combined, leading to a hybrid technique. Industries can keep their IIoT devices running fast and effective without losing valuable data that could help them develop services and drive innovation by using a hybrid technique, i.e., combining the data collection capability of edge computing with the storage capacity, the data handling capacity of fog computing through its ability to process data in realtime, and the processing power of the cloud computing.

The majority of IIoT endusers in the cloud currently have minimal control and are restricted to monitoring and analysis. However, in the future, there might be an IIoT system that can control, monitor, and analyze the process in the cloud.

A major drawback of edge devices is that they only store data collected locally, making it impossible for them to use any sort of predictive Bigdata. Cloud computing facilitates a degree of large-scale data processing at the edge. The cloud can collect large amounts of data with its immense storage and processing capacity and analyze it in a variety of ways to generate great information, patterns, and solutions. Cloud computing's data processing capabilities have also enabled AI and ML to become more feasible. Industries can enhance their ability to incorporate advantages of both methods through integrating edge computing with centralized cloud computing, i.e., fog computing, thus minimizing their limitations.

9.6 CONNECTIVITY ON IIoT

For the IIoT, short-range technologies such as BLE and ZigBee are useful. However, based on the nature of the industry, long-range technologies are also equally important. Mobile cellular communications-based solutions consume more energy, which is sometimes not much useful for IIoT applications.

A low-power wide-area network (LPWAN) is one of the approaches for connectivity that is useful for IIoT. This is a popular technology for many IoT applications. It has up to 40 km communication range in rural areas and 10 km in urban regions;

battery lifetime is up to 10 years, less than $5 of device cost (Centenaro et al., 2016; Patel & Won, 2017). In many IIoT-based applications, it is commonly used because it provides a long transmission range, low energy consumption, simplified network topology (often single hop and star topology), low cost, easy and scalable implementation, thin infrastructure, small frame sizes of data, albeit low data rates. LPWAN technologies usually suffer from problems with network congestion since they are often implemented in the unlicensed ISM band (Onumanyi et al., 2019).

The low-power communication is achieved for IIoT, applying LPWAN with technologies, such as Sigfox, NBIoT, LoRA, Ingenu, and LoRaWAN. Some of these are described below.

a. Sigfox is developed in 2010 by a French global network operator start-up Sigfox that is based on is based in Labège near Toulouse, France. Sigfox operates and commercializes its own IoT solution in more than 70 countries around the world (Sigfox, 2021). Sigfox builds wireless networks to connect low-power IoT objects that require continuously on and emitting small amounts of data. Sigfox employs the differential binary phase-shift keying (DBPSK) and the Gaussian frequency-shift keying (GFSK) that enables communication in the ISM band using 868 MHz in Europe and 902 MHz in the US. It utilizes a wide-reaching signal called Ultra Narrowband that requires little energy, therefore called an LPWAN.

b. LoRa, or Long Range, is a low-power wide-area network modulation technique based on spread spectrum modulation techniques derived from chirp spread spectrum (CSS) technology. It was developed by Grenoble, France based on the start-up Cycleo and acquired by Semtech, USA (Mekki et al., 2018). As its name says, LoRa is targeted for long-range transmissions with low power consumption. It can achieve data rates between 0.3 and 27 kbit/s depending upon the spreading factor. It uses ISM bands like 433, 868 (Europe), 915 (Australia and North America), 865–867 (India), and 923 MHz (Asia). Currently, LoRaWAN is standardized by LoRa-Alliance and is actually deployed in many countries around the world.

c. Narrowband Internet of Things (NB-IoT), also known as LTE Cat NB1, is a narrowband radio technology for IoT applications. The 3rd Generation Partnership Project (3GPP) standardized the NB-IoT. NB-IoT uses a subset of the LTE standard, but the bandwidth is limited to a single 200-kHz narrow-band. It uses OFDM and SC-FDMA for downlink and uplink communications, respectively. Many operators all around the world already deployed NB-IoT.

9.7 CR FOR FUTURE IIoT

9.7.1 Spectrum Scarcity Problem in IIoT and Cognitive-IIoT

As discussed earlier, based on the nature of the industry, in smart factories or smart manufacturing processes, millions of tiny edge devices continuously generate a massive amount of data and transmit data streams, resulting in increased network traffic.

IIoT Edge Network and Spectrum Scarcity Issue

The sensors and actuators and other devices create vast amounts of real-time data, which will need to be processed and analyzed in order to make swift decisions.

Most of the current communication protocols, such as Bluetooth, Wi-Fi, LoWPAN, and ZigBee run in ISM bands that are not sufficient for massive industrial data generated by IIoT devices. Current fixed spectrum utilization policies result in inefficient utilization of spectrum (Joshi et al., 2013). Research has shown that this coexistence in the ISM band can degrade the performance of the WSNs. The wide deployments, large transmit power, and a large coverage range of IEEE 802.11 devices and other proprietary devices can degrade the performance of WSNs significantly when operating in overlapping frequency bands (Joshi et al., 2013).

A cognitive radio network (CRN) provides opportunistic spectrum access with the help of intelligent radio device to overcome spectrum inefficiency. CRN devices sense idle incumbent license bands and access opportunistically by dynamically changing transmitter parameters. This device has cognitive capabilities, reconfiguration, environment sensing, trust and security, and power control capabilities. These devices have a characteristic of fast switching with simultaneous transmissions in addition to channel probing and channel state learning. Dynamic spectrum access (DSA) capability enables a CR user to adapt to varying network conditions. A CR user is allowed to use the spectrum with guaranteed protection to an incumbent license user, also called a primary user (PU).

Cognitive Industrial Internet of Things (Cognitive-IIoT or CR-IIoT) is the use of cognitive computing technologies, which is derived from cognitive science and artificial intelligence, to power next-generation Industrial IoT. One of the main motivations for using CR in IIoT is the allocation of bandwidth for IIoT devices. Because millions of IIoT devices are used in industries, it is difficult to allocate spectrum bands to these devices. The use of CR technology in IIoT will help in the growth of IIoT and solve the spectrum scarcity issues. The current, fixed spectrum allocation policy requires licensing fees to be paid and in many cases spectrum are given by auctioning. The CRNs have two main objectives (i) Utilizing unused licensed spectrum and provide enough bandwidth to the SUs, and (ii) Protect the right to channel access whenever PU wants to access without interference.

The operational planning for Cognitive-IIoT fundamentally characterizes the interactivities of mainly five essential cognitive functions: intelligence cycle, large data test, logic knowledge, derivation, and detection.

Authorities already started standardizing such spectrum sharing policies and encouraging the purpose of the solution to researchers. IEEE 802.22, IEEE 802.11ah, and IEEE 802.15.2 are examples of such standards (Joshi et al., 2015). Researchers and industry are working to improve the performance of Cognitive-IIoT in terms of cost, energy consumption, data rate, robustness, network throughput, QoS, and security. A range of logical techniques has been employed to achieve the required network performance, such as power-aware MAC, cross-layer design technique, efficient sensing technique, and significant enhancement in hardware design (Joshi et al., 2013).

Joshi et al. provided a detailed study on CR-wireless sensor networks in the article – cognitive radio wireless sensor networks framework, applications, challenges, and research trends (Joshi et al., 2013). Figure 9.2 shows the hardware structure of the CR-wireless sensor.

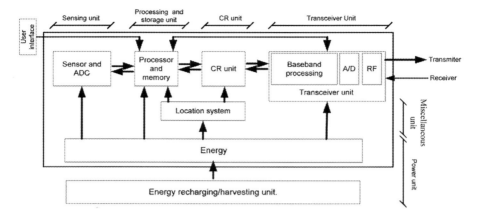

FIGURE 9.2 Design of CR-wireless sensor. (Joshi et al., 2013.)

For seamless communication in Cognitive-IIoT, perfect channel sensing is very important. Basically, there are two techniques for sensing: fast sensing and fine sensing. Fine sensing gives more correct results but consumes more time. Fast sensing, on the other hand, takes less time to detect signals in the channel. The energy detection technique is an example of fast sensing.

This technique can sense the idle channel power of the signal without PU information. In this technique, an ideal channel is determined based on the power of the signal. The output value of energy detection is compared with a predefined threshold value according to the variance of the noise signal. In this technique, determining the threshold value is complicated; therefore, the uncertainty of the noise signal affects the system performance. This technique is based on the Neyman–Pearson approach (Bayram & Gezici, 2011). This approach increases the probability of detection (Pd) depending on false alarm probability. The false alarm (Pf) probability is that the output signal is greater than the threshold value when the primary user is not broadcasting. Let's assume that H0 determines the channel is idle, and H1 determines the channel is occupied by the incumbent user (Atapattu et al., 2011; Yilmazel et al., 2018):

$$X[n] = \begin{cases} W[n], H_0 \\ s[n] + W[n], H_1 \end{cases} n = 1, 2, 3, \ldots \ldots \quad (9.1)$$

where $X[n]$ is a signal received by the SU, $W[n]$ is the signal with zero mean and variance σ_w^2, $S[n]$ is the signal with zero mean and variance, and also is a received signal assuming that it is an independent random process. The signal-to-noise ratio (SNR) can be obtained as follows (Lien et al., 2008):

$$SNR = \frac{\sigma_s^2}{\sigma_w^2} \quad (9.2)$$

IIoT Edge Network and Spectrum Scarcity Issue

Detection probability can be obtained as follows:

$$P_d = Q\left[\frac{\gamma - N\left(\sigma_w^2 + \sigma_s^2\right)}{\sqrt{2N\left(\sigma_w^2 + \sigma_s^2\right)^2}}\right] \quad (9.3)$$

where $Q(.)$ is the general Marcum Q-function, N is the number of samples to be found and can be obtained from the following equation to find Pd and Pf corresponding to a given SNR (Digham et al., 2003):

$$N = 2SNR^{-2}\left[Q^{-1}\left(P_f\right) - \left(1 + SNR\right)Q^{-1}\left(P_d\right)\right]^2 \quad (9.4)$$

The above detection probability is for Awgn Channels. Digham et al. studied an average detection probability for others, such as Rayleigh, Nakagami, and Rician channels (Digham et al., 2003).

After determining the spectrum holes or the idle channel, the channel is accessed via various channel access techniques (Joshi et al., 2014). Along with the PUs' right to access the channel whenever they want, an MAC for Cognitive-IIoT has to consider several aspects such as massive data generated by devices, higher cost, energy conservation, computational complexity, delays, throughput, vertical handoff, heterogeneity, and interoperability, among others.

Cognitive-IIoT aims to provide high performance in communicating, computing, controlling, and a high level of machine intelligence for emerging smart IIoT applications. Cognitive-IIoT covers the relationship among humans, machines, and the surrounding environments. It plays the role of assistant or mentor for the manufacturer and industrial technologist (Hu et al., 2018).

9.7.2 Cognitive LPWAN for IIoT

Cognitive LPWAN is an example of Cognitive-IIoT. In Cognitive LPWAN, the unutilized spectrum is utilized by secondary users via multiple LPWAN technologies as a combination in one backbone network. The network achieves efficient user experiences using AI services and can provide stable communication between users and IIoT devices. The Cognitive LPWAN can be expanded its service area for various applications from smart cities, to the Internet of Vehicles, Internet of Medicine, and finally Internet of everything.

The cognitive devices or SUs need to detect the availability of unused incumbent spectrum using sensing technologies and incumbent license users' arrival predictions. There are several PU arrival prediction algorithms and sensing technologies in the literature (Joshi et al., 2013). SUs access the unused available spectrum band and prepare for other backup channels for handoff in case any PU claims the channel currently utilizing by it.

The advantage of using CR in LPWAN is to use underlined spectrum resources with minimum energy consumption. The Cognitive-IIoT devices are assumed to be

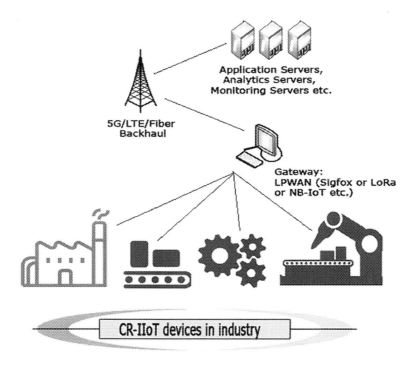

FIGURE 9.3 Cognitive-IIoT architecture.

smart sensors as shown in Figure 9.3, having CR equipped LPWAN base stations responsible for receiving, storing, and processing data and redistributing the requests on the networks. These smart sensors are low-power, low-speed, long-range, LPWAN enabled tiny devices. For long-range and lower-power cellular LPWAN technologies such as LTE, NBIoT is used. A dynamic spectrum access scheme is used in such networks to allow unlicensed users to access unused licensed bands dynamically. Cognitive radio with dynamic spectrum access provides a solution for increasing spectrum utilization in cognitive LPWAN (Rahman et al., 2019). The potentiality of CR-LPWAN is discussed in Rahman et al. (2019). A survey on cognitive radio in low-power wide area networks for industrial IoT applications is presented in Nurelmadina et al. (2021).

CR-LPWAN and other possible variants of Cognitive-IIoT are yet to be standardized, and before that, this area needs extensive study on its effect on spectrum efficiency and right to access of channel of incumbent users.

9.8 CONCLUSION

Edge devices for IIoT are tiny devices with limited computational ability. Data generated by these devices are detailed and massive in amount. Data aggregation and processing is required for intelligent manufacturing and business process. Several

IIoT Edge Network and Spectrum Scarcity Issue

computing schemes such as fog computing, edge computing, and cloud computing are possible. However, based on the industry's requirements, hybrid computing can be carried out.

Connectivity is very important for IIoT, and current connection technologies are, basically, based on ISM bands. Spectrum scarcity issue arises in the ISM band, and it is one of the obstacles in IIoT implementation. There are several solutions to mitigate the spectrum scarcity issue. One very prominent solution is Cognitive-IIoT. One example of Cognitive-IIoT is Cognitive-LPWAN. Cognitive radio technology for IIoT is not standardized yet.

REFERENCES

Atapattu, S., Tellambura, C., & Jiang, H. 2011. *"Spectrum sensing via energy detector in low SNR"*. In: *IEEE International Conference on Communications*. Kyoto, Japan, IEEE, pp. 1–5.

Bayram, S. & Gezici, S. 2011. "On the restricted Neyman-Pearson approach for composite hypothesis-testing in the presence of prior distribution uncertainty". *IEEE Transactions on Signal Processing* 59(10) (irailak): 5056–5065.

Boyes, Hugh et al. 2018. "The industrial internet of things (IIoT): an analysis framework". *Computers in Industry* 101(urriak 1): 1–12. doi:10.1016/j.compind.2018.04.015

Centenaro, M., Vangelista, L., & Zanella, A. 2016. "Long-range communications in unlicensed bands: the rising stars in the IoT and smart city scenarios". *IEEE Wireless Communications* 23(5): 60–67.

Digham, F. F., Alouini, M., and Simon, M. K. 2003. *"On the energy detection of unknown signals over fading channels"*. In: *IEEE International Conference +on Communications, 2003. ICC'03*, 5, pp. 3575–3579, libk.5. doi:10.1109/ICC.2003.1204119

Espinoza, Héctor et al. 2020. "Estimating the impact of the Internet of Things on productivity in Europe". *Heliyon* 6(5) (maiatzak 1): e03935. doi:10.1016/j.heliyon.2020.e03935

Farnell. 2016. "Will there be a dominant IIoT cloud platform?" *Online*. https://uk.farnell.com/will-there-be-a-dominant-iiot-cloud-platform

Hu, L., Tian, D., and Lin, K. 2018. "Cognitive industrial internet of things". *Mobile Networks and Applications* 23(6): 1607–1609.

Joshi, G.P., Acharya, S., and Kim, S.W. 2015. "Fuzzy-logic-based channel selection in IEEE 802.22 WRAN". *Information Systems* 48. doi:10.1016/j.is.2014.05.009

Joshi, G. P., Nam, S. Y., and Kim, S. W. 2013. "Cognitive radio wireless sensor networks: applications challenges and research trends". *Sensors* 13(9): 11196–11228. doi:10.3390/s130911196

Joshi, G.P., Nam, S.Y., and Kim, S.W. 2014. "Decentralized predictive MAC protocol for Ad Hoc cognitive radio networks". *Wireless Personal Communications* 74(2). doi:10.1007/s11277-013-1322-6

Khan, W. Z. et al. 2020. "Industrial internet of things: recent advances, enabling technologies and open challenges". *Computers & Electrical Engineering* 81(urtarrilak 1): 106522. doi:10.1016/j.compeleceng.2019.106522

Lien, S.Y., Tseng, C.C., and Chen, K.C. 2008. *"Carrier sensing based multiple access protocols for cognitive radio networks"*. In: *2008 IEEE international conference on communications*, Beijing, China, pp. 3208–3214.

Mekki, Kais et al. 2018. *"Overview of cellular LPWAN technologies for IoT deployment: Sigfox, LoRaWAN, and NB-IoT"*. In: *2018 IEEE International Conference on Pervasive Computing and Communications Workshops (PerCom Workshops)*, Athens, Greece, IEEE, pp. 197–202.

Microsoft News Center. 2016. "GE and Microsoft partner to bring Predix to Azure, accelerating digital transformation for industrial customers". *Stories*. https://news.microsoft.com/2016/07/11/ge-and-microsoft-partner-to-bring-predix-to-azure-accelerating-digital-transformation-for-industrial-customers/.

Nurelmadina, N. et al. 2021. "A systematic review on cognitive radio in low power wide area network for industrial IoT applications". *Sustainability* 13(1) (urtarrilak): 338. doi:10.3390/su13010338

O'Donovan, P. et al. 2015. "An industrial big data pipeline for data-driven analytics maintenance applications in large-scale smart manufacturing facilities". *Journal of Big Data* 2: 1–26.

Onumanyi, A. J., Abu-Mahfouz, A. M., and Hancke, G. P. 2019. *"Towards cognitive radio in low power wide area network for industrial IoT applications"*. In: *2019 IEEE 17th International Conference on Industrial Informatics (INDIN)*, Helsinki-Espoo, Finland, IEEE, Vol. 1, pp. 947–950.

Paine, T.. 2017. "Three reasons why edge architectures are critical for IIoT". https://internetofthingsagenda.techtarget.com/blog/IoT-Agenda/Three-reasons-why-edge-architectures-are-critical-for-IIoT.

Patel, D. and Won, M. 2017. *"Experimental study on low power wide area networks (LPWAN) for mobile internet of things"*. In: *85th IEEE Vehicular Technology Conference*. Sydney, Australia.

Rahman, M. et al. 2019. *"Implementation of LPWAN over white spaces for practical deployment"*. In: *IoTDI'19: International Conference on Internet-of-Things Design and Implementation*, Montreal, QC, pp. 178–189.

Ranger, S. 2019. "What is the IIoT? Everything you need to know about the Industrial Internet of Things". https://www.zdnet.com/article/what-is-the-iiot-everything-you-need-to-know-about-the-industrial-internet-of-things/.

Siemens PLM Software. 2021. "Industrial machinery catalyst on the cloud". *Siemens Digital Industries Software*. https://www.plm.automation.siemens.com/media/store/en_us/Siemens-PLM-Industrial-Machinery-Catalyst-on-the-Cloud-fs-59876_tcm1023-250316_tcm29-2087.pdf.

Sigfox. 2021. "Our Story - Discover the story behind the making of the Sigfox company and our key milestones". https://www.sigfox.com/en/sigfox-story.

Snyder, S. et al. 2020. "Why organizations are betting on edge computing". *IBM Institute for Business Value* (maiatzak): 1–17.

Tightiz, L. and Yang, H. 2020. "A comprehensive review on iot protocols' features in smart grid communication". *Energies* 13(11) (urtarrilak): 2762. doi:10.3390/en13112762

Yilmazel, R., Seyman, M. N., and Tuna, E. 2018. "Energy detection approach for spectrum sensing in cognitive radio systems". *Uluslararası Mühendislik Araştırma ve Geliştirme Dergisi* 10(2) (ekainak): 1–9. doi:10.29137/umagd.406275

10 Review on Optical Character Recognition-Based Applications of Industrial IoT

Apurva Sonavane and Jimmy Singla

Lovely Professional University, Phagwara, India

CONTENTS

10.1 Introduction 175
10.2 Literature Review 180
10.3 Conclusion..185
References..185

10.1 INTRODUCTION

Optical character recognition (OCR) is a procedure of converting handwritten text from images to computer-decoded text. Research in handwriting recognition is ancient, and researchers are trying to solve this problem since 1914. It is one of the exciting topics in pattern recognition, and its application aims to transform a large number of documents, handwritten or printed into machine-encoded text [1]. Early implementation of OCR can be traced for telegraphy and helping reading devices for the blind. In 1914, Edmund Fournier d'Albe created a portable device that scanned printed pages as moved across it and produced specific tones for different characters and words, known as Optophone. In the late 1920s and 1930s, Emanuel Goldberg created "Statistical Machine" for searching microfilms archives using OCR and patented it in 1931 in the USA; later this patent was acquired by IBM [2].

Handwriting recognition, in general, is classified into two types: offline and online handwriting recognition methods. OCR is an offline process, but nowadays, the online process is also available using APIs. In offline handwriting recognition, automatic conversion of text into letter codes can be used by computer applications. The advantages of offline OCR are that it can be done years after the document is written which makes it more usable in scanning books in the library, postal codes, band checks but cannot be used in real-life applications like self-driving cars to read signboards. Different handwriting styles make it more challenging to implement recognition techniques to get better accuracy. Generally, offline recognition methods involve

DOI: 10.1201/9781003102267-10

acquisition, segmentation, recognition, and postprocessing and output of one stage is input in the next step. But, in the online system, the two-dimensional coordinates of successive points are represented as a function of time and the order of strokes made by the writer are also available [3].

OCR is a research field of pattern recognition, artificial intelligence, and machine vision. In some restricted domains, handwriting recognition is an easy task. For example, in postal code recognition, the problem is reduced due to its input as numeric digits. However, this problem intensifies due to scanning errors, noise, over-lapping numbers, broken characters, and writing mistakes [4]. This problem becomes more daunting when the input space increases from digits to multiple characters to multilingual characters [5].

In the early stages of OCR research, template matching and structural analysis were widely used [6]. Feature matching, also known as feature analysis, is an inter-mediate approach between these two methods. In early methods, these templates were designed artificially or averaged from a few samples. Still, as the samples increased, these methods became inefficient and insufficient to accommodate the variability of large samples and resulted in low accuracy. In the late 1980s, research community turned to learning-based classification methods to take full advantage of artificial neural networks (ANNs) for large data [7].

Several real-life applications, including banking process, document reading required offline handwriting recognition systems. As a result, a lot of research is being conducted to improve its accuracy and performance [8]. While English word's recognition research is done extensively, in this chapter we will discuss many languages, including Arabic, Marathi, Devanagri, Bangla, etc., recognition systems [9] (Figure 10.1).

The demand of automation has benefitted a lot of businesses. There are many mundane tasks which when automated help businesses save a lot of money and increases the productivity of the organization. With OCR in place, many tasks are now automated including data entry for documents, automatic number plate extraction, searching PDFs, etc. [10, 11].

Banking industry has improved a lot of processes by automation including use of OCR. New user signups are processed by software scanning the filled application form and making an entry into the system that is supervised by person. This process speeds up the signup process drastically. Another use is processing checks, which uses the same process and is very helpful both in terms of processing speed and customer satisfaction [12]. The use of OCR in real-life scenarios has forced governments to compete in this field. Many governments have implemented OCR-based applications for number plate detection for both safety and solving crimes. With huge rode networks applications can automatically detect the vehicle number and can create a path chart for it, which is physically impossible for humans to do for all vehicles [13].

The Industrial Internet of Things (IIoT) is the fastest-growing domain which refers to interconnected sensors, instruments, and other devices networked together with computers' industrial applications, including manufacturing as well as energy management. This connectivity permits for data gathering, exchange, analysis, hypothetically enabling progress in productivity and efficiency as well as other economic advantages.

Review on Optical Character Recognition-Based Applications

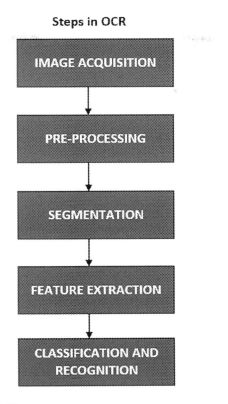

FIGURE 10.1 Steps in OCR.

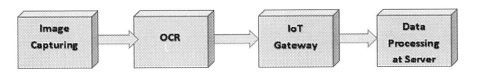

FIGURE 10.2 General steps in IIoT applications using OCR.

There are various applications of IIoT using OCR. It is emerging field in industry to gather data and to process it. OCR plays important role in it. As shown is Figure 10.2. Image acquisition, OCR, IoT gateway and data processing at server level are the common steps in any IIoT applications using OCR.

Steps in General Character Recognition System

Image acquisition: The input picture is taken through camera or other scanners during this process. The picture will be in a common type like JPEG, BMT, etc. The recorded data from scanner or digital camera could be in white, color, or binary.

Pre-processing: Pre-processing is a sequence of operations conducted on the image in the scanned data. Importantly it improves the picture rendering it perfect for segmentation [8].

- De-skew – If the paper was not correctly positioned when checked, it would need to be rotated a few degrees in the clockwise or counter-clockwise direction to match horizontal or vertical lines of text. There are many widely used methods of detecting skew in one page; others focus on identifying linked components and finding the average angles that link their centroids. The skewness will be omitted because this decreases the document's accuracy. The skew angle is measured, and the bent lines are rendered horizontally with the aid of a skew angle [14].
- Noise reduction: There is a high chance that the scanned photos may contain different types of noise (including the one while processing the paper. As a result, these photos with poor quality may affect the processing phase of the next paper. A preprocessing phase is therefore needed to increase the quality of the photographs before they are submitted to subsequent stages of document processing. The broken line section, wide gaps between the lines, etc., may be removed due to the vibration, and it is really necessary to eliminate all such mistakes so that the details can be recovered in the best manner. The pictures include other kinds of noise. One additive noise called "Salt and Pepper Noise," black dots and white dots scattered over a whole picture, usually looks like salt and pepper and can be seen in virtually all papers. Noise reduction strategies may be categorized as extraction, morphological procedures, in two main classes.
    - o Filtering: It aims at reducing noise and decreasing spurious points, typically caused by the data acquisition device's irregular writing surface or low sampling rate. For this reason, specific spatial and frequency domain filters can be built.
    - o Morphological operations: These operations are widely used as an image processing method to isolate image components that are helpful in reflecting and defining the structure of an area. Morphological operations may be used effectively to eliminate the disturbance on the text photos owing to poor paper and ink content, as well as irregular hand activity [15].
- Binarization: Convert a picture from black-and-white to color or greyscale. The binarization function is done as an easy way to isolate the text from the context. Since most commercial recognition algorithms operate only on binary images, the process of binarization itself is important because this tends to be simpler. Furthermore, the efficacy of the binarization process greatly affects the efficiency of the character recognition stage and deliberate decisions are taken in selecting the binarization used for a given type of input picture; as the consistency of the binarization system used to obtain the binary result depends on the type of input picture [16].

Review on Optical Character Recognition-Based Applications

- Thresholding: In order to minimize storage capacity and improve processing speed, it is also preferable to depict gray or color images as binary images by choosing a threshold value for all above that value is set to 1 and all below is set to 0. Two thresholding groups exist, Global and Adaptive. Global thresholding chooses one threshold value for the entire document file, mostly based on an approximation of the background level from the picture's strength histogram. Adaptive thresholding is a tool used for images that might need specific threshold values in various regions of the picture [3, 17].

Segmentation: The Segmentation is the most significant step of Character Recognition techniques. Segmentation is performed to distinguish an image from the actual characters. It is more difficult to divide an unconstrained handwritten word into separate areas (upper, middle and lower) and characters than typed text. This is partly attributable to the variation in the reach, skew, slant, scale and shaped intercharacter as handwriting. Often elements of two consecutive characters may be crossed or overlapped and that condition significantly complicates the role of segmentation. Such contact or combining often happens in Indian languages because of changed characters in the upper and lower areas [18]. There are two types of segmentation, namely External Segmentation [19] and Internal Segmentation [20, 21].

Feature extraction: The extraction of features is the method of extracting the most relevant details from the raw data. The most significant data ensure that the characters will be correctly portrayed on the basis of that. The key goal of extraction of the function is to remove a collection of attributes, optimizing the detection rate for the least number of components. It's challenging because of the complexity of handwriting and its high degree of variation and imprecision in obtaining such functions [17]. Feature extraction methods are based on three types of features:

- Statistical techniques: To some degree depiction of a text picture by statistical point distribution takes care of design variations. While this form of representation does not require the original picture to be replicated, it is used to minimize the measurements of the feature set, to provide high speed and low complexity.
- Structural techniques: The original concept behind the development of structural pattern recognition was the recursive definition of a complicated pattern in terms of simplified patterns dependent on the object's form. These patterns are used to define and identify the CR-systems characters. The characters are depicted as being the union of the primal systems. The basic character derived from writing is believed to be quantifiable, so one may consider the interactions between them.
- Moments: Moments such as central moments, moments in Legendre and moments in Zernike form a closed-packed representation of the original document picture that invariants the process of recognizing an object size, rotation, and translation. Moments are called reflection of series expansion as the initial picture can be recreated and recovered from the coefficients of the moment [22].

Classification: A class mark is applied to the function vector obtained from the preceding step and recognized using supervised and unsupervised process. The data collection for each character is split into a training set and test set. Character classifier can be Bayes classifier, Nearest neighbor classifier, Radial basis function, SVM, Linear discriminant functions, and neural networks that can be implemented with or without backpropagation.

10.2 LITERATURE REVIEW

In Ramteke et al. [5], OCR framework for classification and recognition of Marathi Script is been discussed. In [5], the authors have used WOAR-SVM for classification and new sine cosine algorithm is purposed. They segmented the script in line-segment, word-segment and character-segment with the Modified Pihu method. Various feature extraction was used for extraction of geological and topological from the preprocessed image. They used a self-created dataset with 33 Marathi characters along with 56100 data samples for training and 9900 for testing. With their method (WOAR-SVM), they were able to achieve 95.14% accuracy [5]. However, Lauer et al. [23] presented a new trainable feature extractor based on LeNet5 CNN architecture. They used SVM for classification task to enhance LeNet5 generalization ability. They performed their research on MNIST database and to increase recognition rate affine transformations and elastic distortion were used. They concluded that their method could outperform both SVM and LeNet5. Their method TEF-SVM resulted in 99.17% and 99.46% accuracy without affine transformation and with affine transformation. Du et al. [24] presented a WordRecoder. They presented a unique approach by recognizing the words using acoustic signals generated by paper and pen. They presented an acoustic sensing framework based on new deep learning method, including word suggestion to improve accuracy. Their suggested method was suitable for smart devices and smartwatches. They tested their method with and without the ambient noise with 300-word essay three times each. Their WordRecoder was able to achieve 81% and 75% accuracy for trained and new users.

Tanvir Parvez and Mahmoud [4] in their paper presented an offline Arabic handwriting techniques using structural techniques. They presented a new structural recognition technique while adding an adaptive slant correction algorithm to correct the slant angles for the words. They integrated their purposed structural algorithm in the recognition phase, which is followed by character modeling using fuzzy polygon matching. They performed their work on IFN/ENIT database of Tunisian city names. They concluded that their purposed method is better than single classifier and while being comparable to multi-classifier models resulting in 79.58% word recognition rate.

Kaur and Sagar [9] presented a paper on the Brahmi script, which is an ancient script. They used linear SVM for character recognition. Gradient information of the character images pixels is extracted, and histogram of the gradients is stored as a feature vector for each character image. Their dataset included handwritten character and data taken from the Internet. Feature set of 24 images of each character is used to train linear SVM. Their recognition system is able to achieve 91.6% accuracy. They verified their system using libsvm library for linear svc kernel.

Review on Optical Character Recognition-Based Applications

Jha et al. [25] while also researching on recognition of Arabic text created an Android application for text extraction from captured mobile images. They used the Tesseract OCR library in their work while achieving 90.2% of accuracy.

Pal et al. [26] while working on Bangla handwritten character recognition presented a DBN to create a predictive model to fit both supervised and semi-supervised learning. Their method includes an unsupervised learning, which is followed by supervised fine-tuning of parameters. They used data from the Indian Statistical Institute of basic Bangla characters and numerical Dataset and were able to achieve 90.27% accuracy.

Pal et al. [27] presented a pepper on Devanagri script. In their paper, twelve different classifiers were presented, namely mirror image learning, projection distance, and many more. While this, they used four different feature sets and are computed based on gradient and curvature information. In their paper, they worked on both binary and greyscale images. They collected their dataset from individuals as well as from umapada@isical.ac.in. Here database is divided into five subsets and testing is done on each subset using the rest of the subsets for learning. They averaged the results from each subset to present the accuracy. They presented that they were able to achieve 95.19% accuracy using MIL classifier.

Jha et al. [28] also worked on offline Devanagri script OCR. They used two sets of features combined with two classifiers for better accuracy. The first set of features is determined based on the spatial knowledge derived from the gradient's tangent arc. Finally, they used a combination of SVM and Modified Quadratic Discriminant Function (MQDF). Finally, they were able to achieve 95.13% accuracy on 36172 data samples.

Simon et al. [29] purposed their work for license plate number recognition. They purposed LPRS in which image angles were between −20° and +20°. The rectangular area of the license plate is extracted using edge-based image processing algorithms. Vertical projection is used for character separation. At recognition level, they used three-layer feed-forward neural network using backpropagation network. They tested their system in 357 images and were able to achieve 97% accuracy.

Ram et al. [30] presented new methods for an ensemble-based classifier for English script. They used the hidden Markov Model as a base classifier and performed their experiment on 8795 images of characters and were able to conclude that the method is able to increase the accuracy from 66.23% to 71.58%.

Ram et al. [31] also worked on Devanagari characters. They used DNN for their experiment and optimized their results by selecting the best hyperparameters for their network. While using ReLU activation function with eight layers, they were able to achieve 96.9% accuracy. The highest accuracy for ReLU shows that for this mission, using the nonlinear activation method provides better results.

Binu and Babu [32], in their paper, purposed a new algorithm based on deep learning neural network. They presented that using regularization parameters and suitable activation function improves accuracy. Our suggested CNN model utilizes a variety of convolution layers coupled with activation of ReLU, with dropout being used as a layer of regularization. Dropout regularization is purposed to decrease the overfitting problem. Softmax activation is used at the outer layer, which consists of 10 neurons. This resulted in an accuracy of 97.4%.

Rajashekararadhya and Ranjan [33] while recognizing Malayalam script, presented edge detection in the preprocessing stage. Canny Edge Detector is used to create the character's thinned edges. Before this method, therefore, a technique is applied to concentrate edge regions using nonlinear anisotropic diffusion the partial differential equations (PDEs). Then, these split pieces are related using the process of ant colony optimization (ACO). For the purpose of extracting functionality, this picture is further partitioned into separate areas. Multilayer perceptron (MLP) used these features and rated 95.16% precision of the characters.

Dhanddra et al. [34], in their paper presented a metric-dependent feature extraction method focused on the Zone and Size. The centroid character is determined, and the picture is further broken down into equivalent areas. Average distance to each pixel present in the region is determined from the character centroid. For all the zones present in the numeral picture, this process is replicated. Finally, these characteristics are derived for description and acknowledgement. Feed-forward propagation neural network is intended for the purpose of subsequent detection and identification. We received an identification score of 98% and 96%, respectively, for Kannada and Telugu numerals.

In Chacko et al. [35], handwritten numerals Kannada, Telugu, and Devnagari are considered a method of identification utilizing global and local computational features and a classifier of the Probabilistic Neural Network (PNN). The feature set involves the calculation of spatial density, a system of water reservoir, filled density of the hole, and maximum distance to the image. For Kannada, Telugu, and Devnagari data sets, the overall identification rate was 99.40%, 99.60%, and 98.40%, respectively. It is stated to be free of thinning and without standardization of scale.

Shamim et al. [36] presented a paper on Malayalam character recognition using wavelet energy feature and extreme learning machine. Wavelet transform is used to derive the wavelet energy parameter. Due to slowness of traditional learning algorithms, they purposed ELM for SLFN. They were able to achieve 95.59% accuracy at db6 level and were able to complete the training phase in 212 seconds.

Kamble and Hegadi [37] presented machine-learning algorithms for handwritten character recognition. They obtained their dataset from the Austrian Research Institute for Artificial Intelligence, Austria, which was divided into 1893 training set and 1796 testing set. Their paper presented an offline method using multiple machine learning algorithms and evaluated each method results. After comparison, they were able to conclude that MLP is able to achieve the highest accuracy at 90.37% using WEAK. Each algorithm is evaluated on multiple parameters, including Kappa statistics, mean absolute error, time taken, etc.

Sampath and Gomathi [38] presented a paper on Marathi characters recognition. Rectangle Histogram Based on gradient representation as the basis for character extraction. Their dataset is made up of 8000 samples of 40 simple handwritten Marathi characters each. Among those, 10 examples of each character were obtained from various authors. Both sample photographs of handwritten Marathi characters are uniform to a scale of around 20 pixels. They presented their experimental results on SVM with 95.64% and FFANN with 97.15% accuracy. Results indicate using FFANN results in better output.

Jha et al. [39] purposed a hybrid neural network for English script OCR. They used Charas74K dataset, which consisted of 62 classes and 7705 characters. Before

Review on Optical Character Recognition-Based Applications

resizing the image, they used a median filter to remove the noise from the images. To identify the characters, they purposed an FLM network which is a combination of Firefly and the Levenberg–Marquardt (LM) algorithm and neural network. Finally, they integrated a feed-forward network with their purposed FLM-based neural network and were able to achieve 95% accuracy. In their paper, they compared their results with existing hybrid methods using matrices of accuracy, FRR and FAR.

Soora and Deshpande [40], in their paper, presented a method for OCR of Arabic script. They used IFN/ENIT dataset for their research which consisted of 32492 words. They used machine learning approaches namely k-NN and NN in their paper. Three stages, namely, pre-processing, feature extraction, and classification, are presented in their paper in detail, which concluded that they were able to achieve 80.75% accuracy using k-NN.

Kale et al. [41] presented a paper on OCR of multilingual scripts. Their work included the scripts of Devanagri, English, Bangla, and numerals. They used multiple feature vectors, including cross-count features. They were able to achieve 98.5% accuracy rate on proprietary data-set and media-lab LP benchmark dataset which consisted of 30,00 characters from 280 documents.

Alkhateeb et al. [42] addressed the problem of compound characters in Devanagari scripts. They purposed a system using Zernike moment feature. For handwritten character recognition, the rotation invariance property is desirable at Zernike moment. The picture is pre-processed and compressed to the size of 30 × 30 pixels and further separated into the region, preclassification is performed, and the Zernike function is then extracted from each section. They used 27000 samples from different people and implemented SVM and KNN and found 98.37% and 95.82% accuracy.

Jha et al. [43] also presented a paper on Arabic cursive texts. They purposed an offline word-based recognition system using HMMs. After segmentation, a set of intensity features are extracted which is based on sliding window on mirrored word images. They also extracted structure features and combined for classification. HMM classifier is trained using intensity features and structure features are used for rerank results to increase the recognition rate. They performed their experiment on IFN/ENIT database. They were able to achieve 95.15% and 93.18% with and without reranking.

Su et al. [44] presented IIoT system prospective. This paper proposed the architecture for an IoT device as an automated industrial meter reader that uploads the gathered numeral data to a server for centralized data processing. The implementation of the device is done using Raspberry Pi as the platform. The device follows a four-step process: Image Acquisition using Raspberry Pi camera module, Optical Character Recognition using feature extraction technique, Internet Upload Mechanism using Google Forms, and Online Data Processing using Google Spreadsheet.

Yadav et al. [45] proposed A Finger-Worn Device for Exploring Chinese Printed Text With Using CNN Algorithm on a Micro IoT Processor. In the paper, Chinese OCR system was developed according to the training strategy of an augmented convolution neural network. This system also falls under IIoT domain.

Chaudhuri et al. [46] is another example of IIoT system using OCR. This paper proposed Vehicle Detection application as well as Classification for a Toll Charging System Based on TESSERACT OCR.

Author	Year	Script	Dataset	Feature	Classification Techniques	Results Claimed
[4]	2013	Arabic	IFN/ENIT	Structural features	Fuzzy polygon matching	Accuracy-79.58%
[5]	2019	Marathi	-Self-created Dataset -33 Marathi characters	Geometrical and Topological features	WOAR-SVM	Accuracy-95.14%
[9]	2019	Brahmi	-1290 images -1032 training samples -258 validation samples	Gradient features	Linear SVM	Accuracy-91.6%
[23]	2007	English	MNIST	TEF-based on LeNet5	TFE-SVM	Accuracy-99.46%
[26]	2013	Bangla	-Indian Statistical Institute dataset -27900 images for training and -8600 images for testing	Unsupervised learning approach	DBN	Accuracy-90.27%
[28]	2008	Devanagari	36172 data samples	Gradient features	SVM + MQDF	Accuracy-95.13%
[27]	2009	Devanagari	-Self-created dataset -36172 samples in dataset	Fradient and curvature features	MIL	Accuracy-95.19%
[34]	2008	Kannada and Telugu numerals	-Self-created dataset -2000 Kannada samples -600 Telgu samples	Zone and Distance metric-based feature extraction	Feedforward BPNN	Accuracy-98% (Kannada) 96%(Telgu)
[35]	2010	Kannada, Telugu, and Devnagari	-	-	Probabilistic Neural Network (PNN)	Accuracy- 99.40% (Kannada), 99.60% (Telgu) 98.40% (Devanagri)
[37]	2018	Numeric	-Dataset from Austrian Research Institute for Artificial Intelligence, Austria. -1893 training set -1796 testing set.	-	Multilayer Perceptron	Accuracy-90.37%
[42]	2013	Devanagri	27000 samples	Zernike moment feature	SVM	Accuracy-98.37%
[43]	2011	Arabic	IFN/ENIT	Structural features	HMM	Accuracy-95.15%
[47]	2002	Oriya	Not Mentioned	Structural and Template features	Tree classifier	Accuracy-96.3%
[48]	2000	Gurumukhi	Not Mentioned	Structural and Topological features	Tree classifier	Accuracy-97.3%
[44]	2016	Numerical Data	100 samples	-	-	Accuracy- 65%
[45]	2019	FRT dataset	3000 samples	-	CNN	Accuracy- 96.75%.

10.3 CONCLUSION

In this chapter, we discussed the general OCR system steps and the dire need for practical applications. We reviewed each step in OCR and addressed the advantages and drawbacks of particular methods in their usage. Multilingual scripts including English, Devanagari, Bangle, Oriya, and Arabic recognition system are highlighted. We compared different algorithms on the same language script and reviewed their accuracy and performance matrices. We get to know the challenges faced by researcher related to datasets of many languages. We hope our survey will help the researchers in this research field.

REFERENCES

[1] V. L. Sahu and B. Kubde, "Offline handwritten character recognition techniques using neural network: a review," *Int. J. Sci. Eng. Res.*vol. 1, no. 1–3, pp. 87–94, 2013.

[2] S. Jha et al. "Comparative analysis of time series model and machine testing systems for crime forecasting," *Neural Comput. Appl.*, 33, 10621–10636, 2020.

[3] N. Arica and F. T. Yarman-Vural, "An overview of character recognition focused on off-line handwriting," *IEEE Trans. Syst. Man Cybern. Part C Appl. Rev.* vol. 31, no. 2, pp. 216–233, 2001, doi:10.1109/5326.941845

[4] M. Tanvir Parvez and S. A. Mahmoud, "Arabic handwriting recognition using structural and syntactic pattern attributes," *Pattern Recognit.* vol. 46, no. 1, pp. 141–154, 2013, doi:10.1016/j.patcog.2012.07.012

[5] S. P. Ramteke, A. A. Gurjar, and D. S. Deshmukh, "A novel weighted SVM classifier based on SCA for handwritten Marathi character recognition," *IETE J. Res.* pp. 1–13, 2019, doi:10.1080/03772063.2019.1623093

[6] S. Mori, C. Y. Suen, and K. Yamamoto, "Historical review of OCR research and development," *Proc. IEEE*vol. 80, no. 7, pp. 1029–1058, 1992, doi:10.1109/5.156468

[7] C. L. Liu and H. Fujisawa, "Classification and learning methods for character recognition: advances and remaining problems," *Stud. Comput. Intell.* vol. 90, no. 2008, pp. 139–161, 2008, doi:10.1007/978-3-540-76280-5_6

[8] S. H. J. Pradeep, E. Srinivasan, "Diagonal based feature extraction for handwritten alphabets recognition," *Int. J. Sci. Inf. Technol.* vol. 3, no. 1, pp. 27–38, 2011, doi:10.5121/ijcsit.2011.3103

[9] S. Kaur and B. B. Sagar, "Brahmi character recognition based on SVM (support vector machine) classifier using image gradient features," *J. Discret. Math. Sci. Cryptogr.* vol. 22, no. 8, pp. 1365–1381, 2019, doi:10.1080/09720529.2019.1692445

[10] A. Singh, K. Bacchuwar, and A. Bhasin, "A survey of OCR applications," *Int. J. Mach. Learn. Comput.* vol. 2, no. 3, pp. 314–318, 2012, doi:10.7763/ijmlc.2012.v2.137

[11] R. Kaur, "Text recognition applications for mobile devices," *J. Glob. Res. Comput. Sci.* vol. 1, no. 1, pp. 20–23, 2010.

[12] "Top 5 business uses of OCR technology to help improve user experience". from https://scopicsoftware.com/blog/the-top-5-business-uses-of-ocr/

[13] "OCR applications: automatic data extraction - CVISION technologies." from https://www.klearstack.com/intelligent-data-extraction-invoice

[14] N. Pratap, "A review of Devnagari character recognition from past to future," *Comput. Sci. Telecommun*vol. 3, no. 6, 2012.

[15] S. Mo and J. Mathews, "Adaptive, quadratic preprocessing of document images for binarization," *IEEE Trans. Image Process* vol. 7, no. 7, pp. 992–999, 1998.

[16] J. H. Jang and K. S. Hong, "Binarization of noisy gray-scale character images by thin line modeling," *Pattern Recognit.* vol. 32, no. 5, pp. 743–752, 1999, doi:10.1016/S0031-3203(98)00019-3

[17] O. D. Trier, A. K. Jain, and T. Taxt, "Feature extraction methods for character recognition-a survey," *Pattern Recognit.* vol. 29, no. 4, pp. 641–662, 1996.

[18] W. Niblack, *An introduction to digital image processing.* DNK: Strandberg Publishing Company, Copenhagen, Denmark, 1985.

[19] L. O'Gorman, "The document spectrum for page layout analysis," *IEEE Trans. Pattern Anal. Mach. Intell.* vol. 15, no. 11, pp. 1162–1173, 1993.

[20] S. Randriamasy and L. Vincent, *Benchmarking page segmentation algorithms*, pp. 411–416, 1994, doi:10.1109/CVPR.1994.323859

[21] R. Casey and E. Lecolinet, "A survey of methods and strategies in character segmentation," *Pattern Anal. Mach. Intell. IEEE Trans.* vol. 18, pp. 690–706, 1996, doi:10.1109/34.506792

[22] Y.C.Chim,A.A.Kassim,andY.Ibrahim,"Characterrecognitionusingstatisticalmoments," *Image Vis. Comput.* vol. 17, no. 3–4, pp. 299–307, 1999, doi:10.1016/s0262-8856(98)00110-3

[23] F. Lauer, C. Y. Suen, and G. Bloch, "A trainable feature extractor for handwritten digit recognition," *Pattern Recognit.* vol. 40, no. 6, pp. 1816–1824, 2007, doi:10.1016/j.patcog.2006.10.011

[24] H. Du, P. Li, H. Zhou, W. Gong, G. Luo, and P. Yang, "*WordRecorder: accurate acoustic-based handwriting recognition using deep learning,*" *Proc. - IEEE INFOCOM* vol. 2018, pp. 1448–1456, 2018, doi:10.1109/INFOCOM.2018.8486285

[25] S. Jha et al. "Mitigating and monitoring smart city using internet of things," *CMC* vol. 65, no. 2, pp. 1059–1079, 2020.

[26] U. Pal, T. Wakabayashi, and F. Kimura, "*Comparative study of Devnagari handwritten character recognition using different feature and classifiers,*" *Proc. Int. Conf. Doc. Anal. Recognition, ICDAR*, pp. 1111–1115, 2009, doi:10.1109/ICDAR.2009.244

[27] U. Pal, S. Chanda, T. Wakabayashi, and F. Kimura, "*Accuracy improvement of devnagari character recognition combining SVM and MQDF,*" *ICFHR*, pp. 367–372, 2008.

[28] S. Jha et al. "Recurrent neural network for detecting malware." *Comput. Secur.* vol. 99, p. 102037, 2020.

[29] G. Simon, H. Bunke, and C. Bern, "*Creation of classifier ensembles for handwritten word recognition using feature selection algorithms,*" *Proc. Eighth Int. Workshop Front. Handwriting Recognit.* IEEE, 2002.

[30] S. Ram, S. Gupta, and B. Agarwal, "Devanagri character recognition model using deep convolution neural network," *J. Stat. Manag. Syst.* vol. 21, no. 4, pp. 593–599, 2018, doi:10.1080/09720510.2018.1471264

[31] A. Ashiquzzaman and A. K. Tushar, "*Handwritten arabic numeral recognition using deep learning neural networks,*" *2017 IEEE Int. Conf. Imaging, Vis. Pattern Recognition, icIVPR 2017*, pp. 1–4, 2017, doi:10.1109/ICIVPR.2017.7890866

[32] P. C. Binu and A. P. Babu, "*Pre and post processing approaches in edge detection for character recognition,*" *Proc. - 12th Int. Conf. Front. Handwrit. Recognition, ICFHR 2010*, no. 1, pp. 676–681, 2010, doi:10.1109/ICFHR.2010.111

[33] S. V. Rajashekararadhya and P. V. Ranjan, "*Neural network based handwritten numeral recognition of Kannada and Telugu scripts,*" *IEEE Reg. 10 Annu. Int. Conf. Proceedings/TENCON*, 2008, doi:10.1109/TENCON.2008.4766450

[34] B. Dhanddra, R. G. Benne, and M. Hangarge, "Kannada, Telugu and Devanagari handwritten numeral recognition with probabilistic neural network: a novel approach," *Int. J. Comput. Appl.* vol. 26, 2010, doi:10.5120/3134-4319

[35] B. P. Chacko, V. R. Vimal Krishnan, G. Raju, and P. Babu Anto, "Handwritten character recognition using wavelet energy and extreme learning machine," *Int. J. Mach. Learn. Cybern.* vol. 3, no. 2, pp. 149–161, 2012, doi:10.1007/s13042-011-0049-5

[36] S. M. Shamim, M. B. A. Miah, A. Sarker, M. Rana, and A. Al Jobair, "Handwritten digit recognition using machine learning algorithms," *Indones. J. Sci. Technol.* vol. 3, no. 1, pp. 29–39, 2018, doi:10.17509/ijost.v3i1.10795

[37] P. M. Kamble and R. S. Hegadi, "Handwritten Marathi character recognition using R-HOG feature," *Procedia Comput. Sci.* vol. 45, no. C, pp. 266–274, 2015, doi:10.1016/j.procs.2015.03.137

[38] A. K. Sampath and N. Gomathi, "Handwritten optical character recognition by hybrid neural network training algorithm," *Imaging Sci. J.* vol. 67, no. 7, pp. 359–373, 2019, doi:10.1080/13682199.2019.1661591

[39] S. Jha et al. "Real time object detection and tracking system for video surveillance system," *Multimedia Tools Appl.*, vol. 80, pp. 1–16, 2021.

[40] N. R. Soora and P. S. Deshpande, "Novel geometrical shape feature extraction techniques for multilingual character recognition," *IETE Tech. Rev. (Institution Electron. Telecommun. Eng. India)* vol. 34, no. 6, pp. 612–621, 2017, doi:10.1080/02564602.2016.1229583

[41] K. V. Kale, P. D. Deshmukh, S. V. Chavan, M. M. Kazi, and Y. S. Rode, "*Zernike moment feature extraction for handwritten Devanagari compound character recognition,*" *Proc. 2013 Sci. Inf. Conf. SAI 2013*, pp. 459–466, 2013, doi:10.14569/ijarai.2014.030110

[42] J. H. Alkhateeb, J. Ren, J. Jiang, and H. Al-Muhtaseb, "Offline handwritten Arabic cursive text recognition using Hidden Markov Models and re-ranking," *Pattern Recognit. Lett.* vol. 32, no. 8, pp. 1081–1088, 2011, doi:10.1016/j.patrec.2011.02.006

[43] S. Jha, D. Prashar, and A. A. Elngar. "A novel approach using modified filtering algorithm (MFA) for effective completion of cloud tasks," *J. Intell. Fuzzy Syst.*, 1–9. doi:10.3233/JIFS-189159

[44] Y.-S. Su, C.-H. Chou, Y.-L. Chu, and Z.-Y. Yang. "A finger-worn device for exploring Chinese printed text with using CNN algorithm on a micro IoT processor," *IEEE Access* vol. 7, pp. 116529–116541, 2019, doi:10.1109/access.2019.2936143

[45] B. C. Yadav, S. Merugu, and K. Jain, *Iccce 2018*, Springer, Singapore, vol. 500, no. January, 2019.

[46] B. B. Chaudhuri, U. Pal, and M. Mitra, "Automatic recognition of printed Oriya script," *Sadhana* vol. 27, no. 1, pp. 23–34, 2002, doi:10.1007/BF02703310

[47] G. S. Lehal and C. Singh, "*A Gurmukhi script recognition system,*" *Proc. Int. Conf. Pattern Recognit.* vol. 15, no. 2, pp. 557–560, 2000, doi:10.1109/icpr.2000.906135

11 Using Blockchain in Resolving the Challenges Faced by IIoT

Nishant Jha and Deepak Prashar

Lovely Professional University, Phagwara, India

CONTENTS

11.1 Introduction 189
11.2 What Is a Blockchain? Structure and Concepts 191
11.3 Smart Contracts 195
11.4 Consensus Algorithm 197
11.5 Internet of Things (IoT) and Its Technologies 198
11.6 Trust and Information Security in IoT 200
11.7 Blockchain and IoT..202
11.8 Industrial IoT (IIoT) and Emergence of Industry 4.0 203
11.9 Technological Advancements Related to Industry 4.0...............................204
11.10 Use Cases of IIoT...207
11.11 IIoT Architecture and Scenario 208
11.12 Aims and Limitations of IIoT 209
11.13 Challenges Faced by IIoT and How Blockchain Helps in Resolving
These Challenges 210
11.14 Blockchain in Resolving the Challenges Faced by IIoT............................211
11.15 Conclusion 213
References..213

11.1 INTRODUCTION

According to a forecast [1], with more than 50 billion Internet-connected devices, wireless data traffic will increase by 1000 fold by 2020. Small embedded devices with sensors and actuators would be several of these, allowing new kinds of intelligent applications. All these smart devices are linked and form what we call the Internet of Things (IoT) and collaborate together. We will have unavoidable problems in the future because of the size of devices and the sheer number of sensors, actuators and the information they create if we want to allow safe communication between all these IoT-connected devices. In particular, taking into account the industrial IoT (IIoT), where factories would also have all their equipment linked. In order to prevent tampering attacks and failures, these industries also need to ensure that the integrity of sensor values is maintained at all times [1]. Using blockchains is one new

DOI: 10.4324/9781003102267-11

technology to address these. Blockchains act like a distributed, transparent database consisting of a series of ordered blocks shielded from tampering and revision. With blockchains, the data measured by some industry sensors or actuators can be securely stored in the block without worrying about the quality of the data that can ensure the integrity of the entire industrial system [2]. For the IoT industry, where the vast amount of computer transactions makes for a larger attack vector and confidence problems, the creation and implementation efforts of blockchains tend to be increasingly important. Blockchains are seen here as a ledger that without central storage or control can monitor and verify such computer transactions immutably. However, there has not been a norm set yet, as in the DLT field itself. A broad number of companies and organizations, however, work on solutions and initiatives that incorporate IoT and blockchain [3]. IoT is receiving a lot of attention in industries nowadays. However, security and privacy challenges are increasing due to the lack of fundamental security technologies [4]. The currently available security and privacy methodologies are inappropriate for IoT due to its decentralized topologies and the resource limitations of the mobile devices [5]. Blockchain technology is suggested for ensuring the security of IIoT due to its decentralized nature. It is a type of distributed ledger in which all the blocks are chained together and is able to trace and coordinate with transactions and save that information in IoT [6]. The major advantage of a blockchain is decentralization. It uses the motion time stamping, distributed consensus, encryption of data and economic incentives [4]. It helps in reducing the costs and increasing efficiency and resolves the problem of insecure storage of data in centralized institutions. Big data processing and analysis, the IoT, applications, and services play an important role in the comfort of human everyday life, with recent advanced technology heading towards a hyper-connected society from the growing digital interconnection of humans and items. However, due to the lack of confidence, numerous issues were anticipated that hindered the IoT's growth. In the age of the IoT, trust was thoroughly explored as an extension of the standard triad of protection, privacy and reliability to provide safe, accurate and seamless communications and services. Despite a large amount of trust-related IoT research, however, the prevalent trust definition, models, and assessment and management processes have remained debatable and under development [7].

The IoT is supposed to observe different facets of human life anywhere on Earth with billions of sensing and actuating instruments deployed. Valuable information identifying occurrences and events related to various real-world phenomena is aggregated, processed, and analyzed into observation data. With knowledge from the cyber and social realms, it is possible to expose the untapped operational efficiencies for a variety of software and services and create an end-to-end feedback loop between the needs of individuals and responses to physical items. To do so, it is important to establish a unified CPSS paradigm that "takes a human-centered and holistic view of computing through the study of physical, cyber, and social environment observations, information, and experiences"[8]. Most IoT-related research papers focused on RFID and Wireless Sensor Networks in the early years, aimed at building underlying networking protocols, hardware and software components to allow physical objects and cyberspace to connect and communicate. A human-centered IoT ecosystem, however, in which humans play a significant role in

Using Blockchain in Resolving the Challenges Faced by IIoT 191

supporting applications and services, is becoming increasingly perceptible. This is illustrated by the high rate of use of social phenomena in the production of real-world IoT services and crowd intelligence. It is envisaged that people are an important part of the IoT ecosystem [9, 10]. The merging of physical objects, cyber components, and humans in the IoT, however, would raise new questions about risks, privacy, and security. Consequently, risk management and IoT security are broad in nature and face greater challenges than the conventional triad of honesty, confidentiality, and accessibility in terms of privacy and security [11]. Trust has been recognized in this regard as a significant role to play in helping both people and services to resolve the perception of ambiguity and risk in decision-making. Trust is a multifaceted term that is affected both by participants and environmental factors used in many disciplines in human life. IoT ecosystems are currently based on a riddle of physical objects and networking devices, bundled in a protocol enigma and secured by collections of incoherent structures of protection and privacy. New questions about threats, privacy, and protection at all levels of infrastructure, facilities and society will be posed by the convergence of physical artefacts, cyber components, and, in particular, people. Confidence assessment could therefore mitigate unexpected risks and optimize predictability, allowing both IoT infrastructures and services to function in a managed and autonomous manner and to prevent unforeseen circumstances and service failures [7]. The fourth industrial revolution, also known by the name of Industry 4.0, is characterized as an interdisciplinary phenomenon involving the incorporation of communication and information technology in the context of industrial production and the environment. There is actually a coined word called IIoT under "Industry 4.0," which is one of the dominant innovations that are transforming industries in the modern digital revolution. IIoT is a modern and evolving industrial ecosystem that can be used by companies primarily to improve their production, performance, and reliability. IIoT, along with advanced predictive analytics and intelligent machines, is a blend of machine–human interaction and collaboration. This combination also involves the possibility of gathering real-time data along with quicker and easier delivery of data, allowing the various stakeholders more involved in the end-to-end phase of services inside and outside an organization [12]. This chapter reviews and analyzes the fact of adoption of IIoT and the value generation along with the challenges associated with IIoT. Furthermore, measures to resolve the challenges are reviewed with the adoption of blockchain technology in IIoT and its effects on IIoT.

11.2   WHAT IS A BLOCKCHAIN? STRUCTURE AND CONCEPTS

A blockchain can be interpreted as a distributed database, where all participants are given access to a modified version stored locally. A permanent and irreversible record of all transactions ever made between members of the blockchain network constitutes the database. Cryptocurrencies such as Bitcoin do not directly exchange the value exchanged and registered on the blockchain. Instead, it is possible to pass over a blockchain everything that can be digitally represented. Blockchain technology makes it possible to digitize originality in a modern way that can theoretically minimize the space between our physical and virtual world [13]. The decentralized structure of the peer-to-peer network on which a blockchain is based ensures that all network members have the option of maintaining a local copy of the latest version of the blockchain. At any given time, anybody in the blockchain can check that new as well as old transactions have been carried out correctly and are stored on the blockchain [13]. "When the pseudo-anonymous author Satoshi Nakamoto published the white paper" Bitcoin-A peer to peer digital cash system" [14] in 2008, the idea of the blockchain was discovered. In order to achieve a digital cash system, Nakamoto suggested integrating a blockchain mechanism of organizing data with cryptography and the possibility of decentralizing data. A widely discussed subject today is blockchain technology. It has been announced As the "greatest innovation since the internet" [15]. With this hype, new actors wanting to join the stage. In spite of being an immature field of science, blockchain technology provides fresh and unique possibilities in a virtual world to ensure originality, honesty, and reliability. However, because the science of technology is immature, there have not been any standardization facilities yet. Examples of these are what features a blockchain needs to fulfil in order to be called a blockchain, meanings of terminology and how blockchains can be related to the outside world [13]. The investments in blockchain technology by capitalists rose from 93 million USD to 550 million USD in three years and is predicted to rise to 2.3 billion USD by the year 2021. Blockchain has developed to the future of programmable blockchains aimed to provide a decentralized solution for various IoT and IIoT applications [4].

Blockchain technology is very self-explanatory and is a database generated by a series of data blocks, just as its name implies. The explanation why a chain is considered is that all new data blocks added to the database must refer to the last block already added to the database. This implies that knowledge is in chronological order in the chain [16]. It is considered that all the data in a single block is added at the exact same time [17]. There are four major characteristics of blockchain technology [18].

- Decentralization: Decentralization In comparison to a centralized strategy, there is no need to validate data through a single point that could behave in the device like a bottleneck. Accordingly, third parties are absolutely eliminated for transactions between nodes in a blockchain.
- Persistency: Transactions can be rapidly checked and invalid transactions are less likely with an incentive behind mining. It is almost difficult to erase or alter old transactions, since the block and every single one in the chain behind it will have to be re-validated. Invalid transactions can be quickly found.
- Anonymity: Each user of the blockchain just has to use a randomly generated address instead of using an actual identity. However, there are other ways to recognize a person, both by the design of how the blockchain nodes are treated and by the vulnerabilities of the standard Internet.
- Auditability: For transparency, as seen by the bitcoin blockchain, a system such as Unspent Transaction Output may be used. The blockchain essentially includes transactions that have an unspent status that has already been made and when it applies to one of these, it transfers its status to spent

Using Blockchain in Resolving the Challenges Faced by IIoT

and closes it for further transactions. There is an actual new transaction made. A new unspent transaction is then made instead of using more new transactions. Such unspent/spent then makes transactions easy to check and track.

The architecture of the blockchain database is built on the same basis as the distributed databases, meaning that there is a ledger containing all knowledge that is distributed to all network peers. Both blockchains and distributed databases use consensus models to monitor which data can be added, updated or removed into the database. A distributed ledger ensures that each peer on the network has its own local copy of the current database version, and that all peers receive an updated version of the ledger by the time the data is appended, modified or removed. In order for an intruder to corrupt any data, because of the decentralization, the intruder will have to reach all peers simultaneously instead of just one central group. By nature, this makes decentralization more immune to fraudulent activity [13]. The primary difference between a DBMS and a blockchain is the way information is processed and organized. The arrangement of data in a DBMS is most commonly achieved by establishing various relationship schemes between entities. This means that it is difficult to determine when something was put into the database, since when data first appeared, was updated or removed, there is no built-in timeline. One of the key reasons Bitcoin was so popular is that it provided a solution without relying on a third party to be able to pass value (bitcoins). The structure of how data is appended and structured on the distributed ledger, using blocks instead of relational entity schemas, was one of the components that made this possible [13]. The structure of a blockchain is shown in Figure 11.1.

A blockchain structure consists of connected blocks, where transactions are stored in each block. Whenever a new transaction occurs, the transaction is included with a group of other transactions in the block. This block is added to the blockchain later on, connecting it to the previous block. The first block, the genesis block, must be present in all blockchains. Inside the blockchain client, the genesis block is statically encoded in a way that it can never be altered [20]. The blockchain doesn't contain any data at this point. Instead, it now has an anchor to which it is possible to add new blocks. It will reference the previous block when a new block is added to the blockchain, providing the linkage of a chain. The header of the block consists of metadata that provides information specific to the current block. Just what metadata is included in the block header depends on the architecture of the blockchain. For all blockchains, it is normal for details to be included about the double hash of the block header of the previous block. This ties the blocks to the previous block and thus establishes the structure of the chain. With this characteristic, a blockchain user can monitor at any time that each block consists of the digital signature of the previous block, and therefore relies on the assumption that the data in that particular block has not been changed. Timestamp and block height number, along with a nonce that can be modified in order to fulfill the consensus model, are other details usually included in a block header. The basic information and its format may vary, often depending on what consensus model is applied, between different blockchain designs. Most blockchains contain information in the block header about the Merkle root. The Merkle root is the product of a Merkle tree, based on all transactions included in the payload

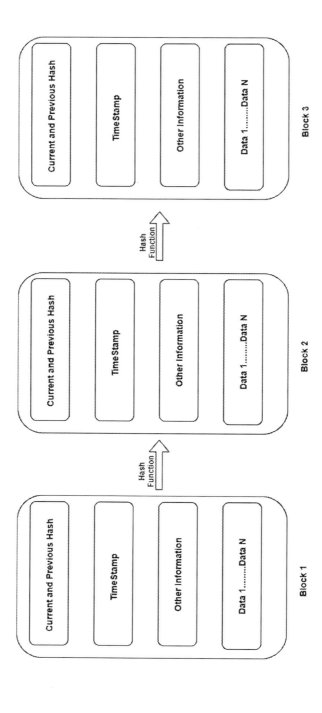

FIGURE 11.1 Structure of blockchain [19].

Using Blockchain in Resolving the Challenges Faced by IIoT   **195**

of the block. It is an important way of ensuring that the blockchain truly contains all transactions contained in the payload. Since every user can derive the Merkle root itself from all transactions, the user can be confident that all transactions are right if this Merkle root is equal to the one in the block header [13].

11.3 SMART CONTRACTS

Smart Contracts are decentralized scripts on a blockchain or relatable infrastructures which allows the execution of predefined processes in a transparent manner [21]. The development of smart contracts dates back to 1994, described as an automated system for executing contract terms via an automated/agreed protocol, such as payment transactions [22]. Therefore, since contract terms are implemented on the basis of pre-identified consensus processes, a typical trusted third party is not necessary [23]. Peters and Panayi [24], meanwhile, suggested a detailed concept of a smart contract: a platform for the compliance and monitoring of trustworthy source data to be stored in BCN on the basis of pre-identified terms of the contract. This is one of the features of the blockchain and a consequence of the recent decade's emerging BCN-ability to transfer cryptocurrency/data across blockchain [25]. According to Andoni et al. [26], smart contracts use peer-to-peer (PTP) networks that allow multi-trusted parties to manage data simultaneously, so that each BCN chain carries its own data and all data is then stored in the ledger in accordance with the accepted consensus mechanism [27]. In addition, smart contracts decrease the reliance on lawyers/third parties to conduct and track contract terms such as financial transactions, thereby improving the consistency and accountability of data [28]. The pseudocode to create a smart contract using solidity [29] is shown below.

PSEUDOCODE FOR CREATION OF SMART CONTRACT (ADAPTED FROM [29])

```
pragma solidity >= 0.4.16 < 0.7.0;

// Defining a contract
contract Storage
{

      // Declaring state variables
      uint public setData;

      function set(uint x) public
      {
              setData = x;
      }
      function get()
        public view returns (uint) {
              return setData;
      }
}
```

Though smart contracts offer a wider range of advantages but there are some challenges also associated with it. It cannot be activated on its own. It gets activated only when it is called in a transaction explicitly. Smart contracts are not automated, instead they require a transaction to activate them [21]. Another challenge is related to the processing of information which does not happen on the blockchain directly. Since a blockchain provides a trusted environment but the information that is transmitted into the blockchain may not be trusted by the participants of the smart contract. This is also called the oracle problem [21]. Every node of the blockchain executes the contract every time, which means that smart contracts are executed globally in the public blockchain. As computation is a deterministic procedure, global execution might not be required [30]. Another challenge is that the data stored is not deleted easily, which conflicts with the legal requirements [31, 32]. It should be noted that the goals of smart contracts, namely, visibility, online enforceability, verification and privacy, result in two distinct directions. In order to limit openness to third actors, secrecy exerts a governing power over the contracts. The other three priorities, visibility, enforceability and verifiability, which require access to contractual data to be passed on to participants or auditors, are diametrically opposed. Therefore, when as little knowledge and control as possible is provided to external parties, an optimum must be sought, but the possibility of verification, observation and compliance is still open. Szabo's solution to the optimization problem in 1997, before blockchain technology and developments in zero-knowledge evidence as well as safe multi-party computations, was to trust an intermediary, a third party, such as an auditor [33].

The Ethereum platform is a general blockchain for smart contracts with a virtual machine (Ethereum Virtual Machine, EVM). Because the environment only operates in the form of a virtual machine on the blockchain, the smart contracts on the node machines are fully disconnected from the network, file system, or other processes. In order to write smart contracts with Ethereum, a high-level Turing-complete language was developed. However, the language, Solidity, has now already become standard for other platforms with smart contract capabilities. In the (relatively) short period the Solidity has been in use, there are a few programming best practices that are unique to the creation of smart contracts [33]. Popular safety guidelines collected during the (relatively) short period of time that solidity was used [33] are as follows:

- Damage control: The amount of tokens stored in a smart contract should be restricted if possible, as the tokens can be trapped in the contract if the source code, platform or compiler contains an error.
- Modularity: It is important to keep smart contracts as small and easy as possible. To keep the contracts as readable as possible, local variables and duration of functions should be limited. The more modular the contracts are, the simpler it is to improve a smart contract structure.
- Checks-effects: At the first step of the algorithm, functions can perform precondition checks. Then, as a second step, state-variable adjustments should be made. Interactions with other contracts could eventually occur.

Using Blockchain in Resolving the Challenges Faced by IIoT

11.4 CONSENSUS ALGORITHM

Proof of Work is the technique used in the Bitcoin network for consensus. A decentralized network also needs someone who tracks transactions and random selection will be a favored technique. But this could contribute to the selection of malicious actors from random drawings. Instead, an artificial workload needs to be overcome so that anyone who wishes to add blocks to the chain could work their way through. In the case of Bitcoin, the workload is usually measured. A hash that is generated from the collective values in the block header is the value that should be measured [34]. If a node has measured one of these valid hashes, it will send the block to all the other nodes with the correct nonce value and the hash calculation is checked collectively and it is added to the blockchain if the given solution checks out on the other nodes as well. Almost at the same moment, with the decentralized existence of a blockchain that would use this technique, two right hashes could be found. This creates two blockchain branches and nodes will continue to run on each of these branches, and then the branch that first gets a new validated block is selected as the real blockchain, while the alternative branch is discarded, allowing the transactions in the discarded branch to have to be re-validated in the new blockchain. The design of this approach implies that most nodes attempt to calculate that most of their resources are spent calculating completely worthless values that effectively waste power [18]. Evidence of participation aims to reduce the energy waste of the proof of work algorithm. This is a focused crypto-currency algorithm that allows them to prove themselves by having a ton of currency instead of having nodes prove themselves by doing calculations. The theory is that a node with a lot invested in the blockchain does not want to harm it and therefore destroy the value of the currency. As the node with the most currency also has the best probability of making more of it, this would end up in an unjust rich get richer scenario. To combat this, options have been built that combine the currency with something else to decide the node that will create the next block. Black coin uses a randomized approach in which a hash value that is generated for each of the nodes, along with the amount of currency that the node is needed to decide who gets the block added [34].

Practical Byzantine fault tolerance is an algorithm that makes it possible for one-third of the nodes not to agree with the rest, but still to send through a block. In what is called a round, blocks are generated where a primary node is selected based on certain rules collection. Then this primary node is in charge of the block formation. Three stages have to be completed to create a block where two-thirds of the nodes have to vote it through each time. Since this solution is based on a specific number of nodes, all nodes in the network must be identified in order for it to function [18]. Ripple uses an approach inside the larger blockchain network where collectively trusted subnetworks are built. The nodes here can be either a client or a server. Servers function like a subnet's main node and manage the validation of new blocks when passing funds only to clients. Servers have a specific list of nodes that are used to decide which nodes a validation request should be sent to. The server adds it to the chain if 80% of the nodes in the list agree that the transaction should be added. Both servers in the wider network share the blockchain [18]. Different consensus mechanism is summarized in Table 11.1

TABLE 11.1

Summary of Different Consensus Algorithms (Adapted from [33])

Consensus Algorithm	Resources Used	Advantages	Disadvantages	Example
POW (Proof-of-work)	Computing power	Trustless, Immutable and highly decentralized	Energy consumption, transaction throughput	Bitcoin, Litecoin.
POS (Proof-of-stake)	Ownership of fixed amount of tokens	Efficient in energy and throughput, scalable	Nothing-at-Stake problem, i.e., voting for different forks at the same time	NXT
Proof-of-Authority (PoA)	Selected authorities are randomly selected to validate transactions	Efficient, doesn't require any inherent tokens or economic value	The corruption of authorities is a large possibility, relies on authorities being well-selected and controlling eachother	Parity POA
PoV(Proof-of-validation)	Security deposit of scarce tokens subject to burn if voting dishonestly	Gives the benefits of proof-of stake without almost any of its draw-backs	Nothing-at-stake problem still persists over long periods of time	Eris-Db
Delegated PoS	Ownership of scarce tokens and peer reputation	Allegedly more efficient than PoS	Voter apathy in elections can lead to excessive centralization and reduced robustness	BitShares

11.5 INTERNET OF THINGS (IoT) AND ITS TECHNOLOGIES

The IoT is a network that links and shares information and useful data with billions of devices, people, and services. In order to improve the level of comfort, productivity and automation for the consumer, IoT applications are highly assertive. The implementation of the IoT automated world includes a high degree of protection and privacy, authentication and recovery from attacks [35]. A member of the Radio Frequency Identification (RFID) development group first proposed the concept of IoT in 1999. Because of the exponential growth of mobile devices, networking, cloud computing, and data analytics, IoT has become more important to the world. More than seven billion people now use the Internet for various types of activities, such as sending and receiving emails, exchanging social media information, reading books, playing games, searching, and shopping online. This wide-scale use of the Internet allows new patterns to be implemented, and this global networking system allows machines to connect and make decisions with each other [35]. Every day, the number of connected devices in the IoT world is rising. The explanation for this rapid increase

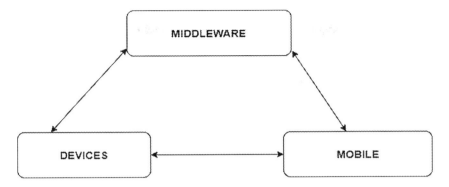

FIGURE 11.2 Communication components of IoT. (Adapted from [35].)

is that connected devices provide comfort and produce good results compared to humans, explained by Burhan, Rehman, Khan and Kim [36]. IoT apps minimize human efforts because they automatically execute tasks In addition to the advantages of these systems, they still have to face challenges, protection and privacy are one of the biggest challenges. The most critical aspect of the IoT is connectivity, since all connected devices have to be able to communicate with each other [35].

Figure 11.2 displays the main IoT components for communication. (a) Hardware: consists of sensors, actuators, physical elements, etc. (b) Middleware: Which is used for data storage and includes programming software used for data analysis; and (c) Presentation: software for visualization and description that can be accessed widely on various platforms [35]. Interconnected devices will be improved by the growing capabilities of various technologies such as RFID, Wireless Sensor Network (WSNs) and increased storage capability of these technologies. The various items of our everyday life, such as people, cars, computers, books, television sets, mobile phones, clothing, food, medicine, passports, luggage, etc., will have at least one unique identity that will enable them to communicate with each other [37]. IoT is used to link various products with the digital world, with the advent of technology such as sensors, smartphones, cloud computing, communication capabilities, etc.[37], this interconnection between devices is increasing. The IoT is a network of various physical objects that use different technology to share data on the Internet, such as cars, computers, home appliances, and more. Technologies that endorse the IoT principle are summarized in Table 11.2.

Identification technologies: Connected devices need to be identified uniquely in the IoT world. For the specific identification of connected devices, identification technologies such as RFID and WSN are used. Network and communication technologies: technologies such as Global Mobile Communication System (GSM), Universal Mobile Telecommunications (UMTS), Wireless Fidelity (Wi-Fi), Bluetooth, and ZigBee allow devices to communicate with each other. In order for the user to use the network with complete confidence and security assurance, communication between the connected devices must be secure. Software and hardware technologies: Smart devices with high device-to-device connectivity can result in smart systems offering high levels of intelligence and autonomy, enabling the rapid

TABLE 11.2
Various IoT Technologies (Adapted from [35])

IoT Technology	Supported Technology
Identification technology	RFID, WSN
Networks and Communication technology	GSM, UMTS, Wi-Fi, Bluetooth, ZigBee
Software and Hardware technology	Smart devices with enhanced inter-device communication

implementation of IoT applications [37]. Various IoT applications are now emerging in order to make our lives more comfortable and effective. A survey conducted by the early adopters of IoT in the UK, US, Japan and Germany by KRC Research has revealed what kind of system is most common among customers. The response we can find is that smart apps have really changed our lifestyle. Some of the applications are listed below [1].

- Smart house: Smart house allows all electronic products, communications products and home information appliances to communicate and share data, achieving the interconnection between different types of electronic products in the home network and achieving control of smart devices anytime, anywhere. There's no need for us to turn to electronic goods when we leave home. The smart house itself regulates temperature and humidity, which invisibly decreases family expenses and makes our lives simpler and more comfortable.
- Safe City Construction: Use the sensors installed on the streets to conduct intelligent image-sensitive analysis and communicate with the police to understand the relation between probes and probes, probes and people, probes and alarm systems in order to establish a harmonious and stable urban living setting.
- Food safety: For each animal on the farm, a two-dimensional code is attached. On the meat sold by the supermarket, the two-dimensional code will be kept. They will first read the two-dimensional code via the cell phone when customers want to buy the meat, and know the history of livestock growth to verify if the meat is safe enough to eat to preserve food safety. In China, this rapid response code already exists for one billion animals.

11.6 TRUST AND INFORMATION SECURITY IN IoT

By incorporating a variety of technologies in many research areas, from embedded systems, wireless sensor networks, service platforms, and automation to safety, security and trust, the IoT is expected to allow advanced services and apps such as smart home, smart grid or smart city. The IoT-related software and services are playing an increasingly important role in the comfort of human everyday life, with

Using Blockchain in Resolving the Challenges Faced by IIoT

recent advanced technologies heading towards a hyper-connected society from the increasing digital interconnection of people and artefacts, Big Data processing and analysis. However, due to the lack of trust that will delay the production of the IoT, there are several problems. It is important to create a trusted and secure environment in order to exchange information, create awareness and conduct transactions in order to cope with a large number of complex IoT applications and services [7]. For organizations and individuals using the information system, information management is an essential part of life. These systems store and exchange critical information that needs protection against a variety of threats that require a number of security controls. It is important to protect certain systems and information from unauthorized access, disclosure, interruption, alteration. Vashi et al. [38] address that the use of IoT is increasingly growing, creating more vulnerabilities and security issues. Burg, Chattopadhyay and Lam [39] demonstrate that a huge wireless and wired infrastructure providing connectivity between devices provides IoT communication and security. The Internet is the base of the IoT, and all of these technologies face the same kind of security problems. The awareness layer, transport layer and device layer are composed of three main layers of IoT. Each of these layers has security issues of its own. Different information security priorities are discussed in Table 11.3 and Table 11.4, respectively.

The IoT is a layered design, consisting of three layers or five layers. The perception layer, network layer and application layer are three layers, and the perception

TABLE 11.3
Objectives of Information Security (Adapted from [35])

Objectives	Explanation
Confidentiality	Confidentiality ensures the information cannot be made public or made available to unauthorized entities.
Integrity	Integrity implies that no one can make adjustments without permission, ensuring accuracy and reliability.
Availability	Availability means that, when needed, data or information should be available.

TABLE 11.4
Objectives of Information Security (Contd.) (Adapted from [35])

Authenticity	Authenticity Implies that Data/Information can be Checked and Trusted and is Authentic.
Accountability	Accountability refers to non-repudiation, deterrence, isolation of errors, identification and prevention of intrusion, and legal action.
Non-repudiation	The evidence of sending and receiving the data is given by both the sender and the recipient.
Reliability	Reliability means that the findings are coherent and as expected.

layer, network layer, application layer, middleware layer and company layer are five layers. Each layer is susceptible to attacks and threats to security. These can be active, or they may be passive. Such threats can come from external sources or from an internal network [35]. Firstly, perception layer attacks could be sensitive information leakage, Denial of Service (Dos) attack, etc. Secondly, network layer attacks may be sybil assault, sinkhole assault, middle attack guy, etc. Finally, malicious code injection, sniffing attachment, etc., may be an attack on the application layer.

11.7 BLOCKCHAIN AND IoT

As summarized in [34], many of the solutions used the bitcoin blockchain back then, which was described as highly inappropriate for IoT applications due to its 10-minute intervals between block writes and scalability issues due to the size of the blockchain. Therefore, latency, blockchain storage problems and how to circumvent proof of work cryptographic demand constraints for devices were the initial list of major issues heading into this examination. Scalability and stability for these completely heterogeneous structures were the two main initial problems within the IoT that the blockchain was supposed to solve. It would not only be perfect for IoT itself if these problems were solved, but it could theoretically also be used as an inspiration for other heterogeneous IT systems [34]. Various solutions such as blockchain servers, PKI reworks, have been suggested in [34] to resolve the challenges which occur in IoT. A new type of blockchain called the redactable consortium blockchain is defined by Du et al. [40]. It allows for blockchain changes that are useful if errors are inserted into the chain or if, for whatever reason, information needs to be modified in general. This new form of blockchain designed specifically for use with power-restricted IoT devices proposes a new hash and signature as building blocks.

A network of chains created by a chain manager. Arranges memberships for all nodes and runs network-building smart agreements. The chain manager is no longer required once a block is added to the blockchain and gives control of the blockchain to the approved nodes and goes offline. The user sensor is a tool that uses a newly developed computer. The transaction authentication algorithm gives approved nodes the go-ahead to update signatures. Registered nodes are permitted by all working together [34] to write and redact in the blockchain. Redaction can be achieved without creating chain forks and signatures can be changed without the need for the user sensor during the redaction. In the event of conflicts occurring during transactions, Judge sensors are there and can distinguish between signatures from user sensors and approved nodes that help settle who is in the right [40]. To avoid the need for calculations to authenticate, nodes share a key used for authentication. For blocks that are best suited for use with low power edge computers, new hashing algorithms. Thanks to blockchain protection, it can be trusted with features that allow blocks to be redacted based on all participating nodes. Uses smart contracts to manage requests for authentication [40]. Kim [41] tries to solve the issue of proof of work by requiring a lot of computational power by using a reverse sequence hashing algorithm to construct blocks. A creator node that generates and holds a private blockchain and also

Using Blockchain in Resolving the Challenges Faced by IIoT 203

mines the new blocks consists of a physical device. It has a manager node that handles edge device authentication and authorization. Edge nodes are called member nodes and a manager node needs to be invited into the network. It can start accessing resources on the network only when a block is written that indicates a member has been authenticated.

Brandão [42] recommends industrial control systems as a system. It has the unusual solution of having a blockchain size set to max and what can be called a rolling blockchain where the first blocks are removed when adding new ones. In the scheme, the only non-edge device is called a pub and is only used to authenticate an edge device into the system initially. The pub has a predefined set of codes generated by the administrator of the system. They will need to send one of these codes in the list to the pub for authentication to authenticate the edge unit. Once the edge computer is authenticated, the pub maintains a list of authenticated devices. The blockchain is then synchronized from all the pubs that are connected with the edge device to the edge device. Each node must be capable of holding a pair of private and public keys that are used for message identification, authentication and encryption. As most communication is device-to-device, they will need to be able to hold the blockchain. As the standard new block pointing to the header of the old block system still exists, at least one message needs to be kept in the blockchain at all times.

11.8 INDUSTRIAL IoT (IIoT) AND EMERGENCE OF INDUSTRY 4.0

Currently, there is a coined word called IIoT within Industry 4.0, which is one of the dominant technologies that are transforming industries in the latest digital transformation (DT). The modern and evolving industrial ecosystem that companies can use to improve their productivity, performance and reliability is the IIoT. IIoT, along with sophisticated predictive analytics and intelligent machines, is a mix of machine–human interaction and collaboration. This combination also involves the possibility of gathering real-time data along with quicker and simpler data delivery, allowing the various stakeholders more involved in the end-to-end phase of services and goods inside and outside an organization [43]. In terms of what the future value creation of introducing and embracing IIoT might be, the advancement of digital technologies such as IIoT has provided industries and manufacturing-focused businesses with the ability to enhance their operations with new technology. There is a lack of research into how to increase and generate value for IIoT and whether the targeted companies are prepared for their DT operations and whether these companies are prepared to implement new and emerging technologies such as IIoT [43]. This process can be streamlined and decentralized in today's sense of Industry 4.0, bringing decision-making to a new era. The expectation is that innovations from Industry 4.0 lead to decision-making that is quicker and more relevant. In order to improve productivity and quality in manufacturing firms, it is clear that the decision-making process for improved business results is being revisited. For instance, how decision-making impacts clients and creates value. This leads to shifts in 'fitting' business models as a result of the technologies of Industry 4.0 [43].

Industry 4.0 has allowed new ways of value creation to affect organizational structures, processes and routines. Traditional backward-looking hierarchical institutions of the day, the systems and routines adopted were naturally spread alongside hierarchies in which a strict top-down flow was information and decision-making. Routines and capabilities have emerged with each functional region as non-immutable and organically breaded. In the sense of Industry 4.0, however, innovations have breached strict hierarchies where contracts are becoming intelligent contracts and automated decision-making is now decision-making. Furthermore, innovations have made it possible to allow new entrants to build new eco-systems for systems. Two points can be defined as the key difference between IIoT and IoT: one is the different focus, the main focus of IIoT is the application of manufacturing and services aimed at producing higher-value equipment and assets such as electricity, transport and industrial control. IIoT actually contains projects with specific market advantages. At the same time, operational safety standards such as data integrity are higher, while IoT pays more attention to the consumption sector, such as home apps, smartphones, etc. [1].

Product efficiency, sustainability and just-in-time development are one of organizations' main concerns at the moment. To solve these problems, lean methods have been used and it is necessary to adopt them with consistency and knowledge for the organizations to succeed. Many companies have been successful in adopting most of the lean methods. The organizations, however, are still lagging behind in bringing their fruit to the fullest. It is a new paradigm, also known as the Fourth Industrial Revolution, also known as Industry 4.0 comes into play [44]. With the application of future-oriented machinery and a state-of-the-art communication and information system [45], Industry 4.0 has been envisaged as a smart factory. This revolution would make the industry more competitive and competitive in manufacturing, while keeping contact a crucial part of it [46]. As stated, however, by Sanders et al. [45], the whole process of introducing and operationalizing I4.0 is a cost-intensive process. As was the case with previous revolutions, it will take some time for countries and businesses to accept the new revolution. Industry 4.0 is a recent concept that is still being studied and investigated by scholars. In addition, search for processes in various industries to incorporate it. At the same time, some studies also suggest that it is too early to worry about its application, and this phenomenon will take ten or more years to fully understand. In addition, the authors note that this notion is far from being known because there are many problems that come with this revolution and have not yet been found out. Those challenges include political, technical, social, economic and science issues [44].

11.9 TECHNOLOGICAL ADVANCEMENTS RELATED TO INDUSTRY 4.0

Industry 4.0, which is based on multiple technologies, is a complicated but adaptable system. It is important to note that these technologies are technologies that are digitally oriented. In addition, the system automates the entire development process and collects real-time data, which can be used for analysis and well-informed decisions by management. Nine innovations form the basis of Industry 4.0, although they are already in use by numerous manufacturing firms. With Industry 4.0, however, these

TABLE 11.5
Description of Various Technologies Advancing with Industry 4.0 (Adapted from [44])

Technology	Explanation
Big Data and analytics	Based on large data sets recorded from several different sources, analytical technology is used to assess the threat, response, prevention, monitor and forecast new issues.
Autonomous robots	Industrial robots that can intelligently complete tasks, with a focus on security, flexibility, versatility, and collaboration.
Simulation	To exploit the real-time data and model the physical manufacturing environment, simulation software is used. This enables an engineer to digitally test, evaluate and refine the environment before any practical change is carried out.
Augmented reality	A real-time view of a real-world physical environment that has been changed or enhanced by superimposing it with virtual computer-generated data. Displays, input devices, monitoring, and computers are the key components of AR technology.
Horizontal and vertical integration	The development of a common and unified data network infrastructure allows for the incorporation and linking of various enterprises, departments and functions, allowing seamless collaboration and an integrated value chain feasible.
Cybersecurity	The provision of secure communications, advanced identity and access control for cybersecurity threat management systems.
Industrial Internet of Things (IIoT)	Networking of various objects embedded with sensors, actuators or other digital data collection and exchange (information) devices. This allows the devices, as appropriate, to communicate and interact with each other and with a more centralized controller. It also decentralizes analytics and decision-making, enabling answers in real-time.
The Cloud	Cloud computing enables data to be exchanged within milliseconds or faster from connected devices to the same cloud. This means that, with the help of cloud systems in real time, cyber-physical systems operating in the manufacturing system can be intelligently connected. Via visualized and scalable resources over the Internet, cloud computing facilitates the distribution of computer services such as servers, storage, databases, networking, apps, analytics and more applications.
Additive manufacturing	Additive manufacturing allowed use of a virtual model, such as complex 3D CAD model data, to manufacture a product through 3D printing or the use of similar technology in a fully automated process.

innovations for the production process are unified. This unification, however, strengthens and automates the development process [44]. Various Technologies are described in Table 11.5.

Ever since the Internet became a reality, computer interconnections have become a certainty as well. It has altered the way people used to get about their everyday lives, has made it easier for people to connect, and gives a sense of reduced distances. Similarly, through its activities, the Internet has even reconditioned the industrial environment. There are smart devices nowadays that are able to perform all the tasks imaginable, and they are hand-held devices. The manufacturing industry's priority is to make full use of these devices and build an intelligent network within organizations [47]. The IIoT is a term that is used in Industry 4.0 for organizational communication, both within and with its stakeholders. IoT's primary purpose is to gather information that it gathers through various networks and sensors installed in the production process. This includes sensing devices such as RFID, infrared sensors, positioning sensors, laser sensors and many other technical devices that are connectable to the Internet for the use of various technologies [48]. Many companies already use sensors and computers in their systems, but most have not incorporated them into their systems. Which means that the elements do not interact with each other. With IoT, however, the devices are interconnected and communicate through a secure wireless link with each other. It makes it simpler, via a centralized configuration, to monitor the machines. However, as per the specified method, the data collection is conducted in a decentralized manner. On the basis of singular processes, this helps in quick decision making. In addition, it enables decisions to be made in real time, thereby making the process more seamless [44]. Another interesting point to note that In fact, the IIoT has no architecture. For the IIoT, the researchers have proposed various architectures such as the three-layer architectures & five-layer architecture [49] as follows.

- Three-Level Architecture: During the early days of IoT science, this design was put into operation. Application Layer: This layer is intended to provide an end user with the relevant features. application layer: numerous applications can be created. Definition of this layer that is used for a particular application, such as a smartwatch, smartphone, smart TV, etc. Network layer: In IoT design, the network layer plays a major role. It connects with other (smartwatches, servers, and more) mobile electronic devices. Using the network layer, sensor data is transmitted and analyzed. Perception Layer: By sensing the atmosphere using sensors, it is the external layer that gathers ambient data.
- Five-Level Architecture: The academics working on IoT have introduced another term that is five-level IoT architecture. The three levels, the system, network, and perception layers, have the same design as the three-level IoT design in this five-level IoT model. The newest level is also the level of industry, level of logistics, and level of computing. Transport Level: This level transmits data from the processing level to the perception level and vice versa through wireless networks such as Local Area Network, Bluetooth, etc. Layer Computing: It is located at the core of the IoT architecture. In this five-level IoT architecture, it has a significant part to play. This stores the documents and processes the data coming from the stage of transport. The new developments such as big data, cloud computing, etc., will be extended to the computing level. Business Level: This level is the IoT Design Master. This governs the entire IoT software, such as applications, client security, model advantages, etc. Both devices connected to a network and interacting with the Internet make up the smart home. The best one to build is a smart system like this five-level IoT protocol stack.

11.10   USE CASES OF IIoT

- Big Data: In any method, big data means the collection of real-time data generated by the sensors. Analytics, on the other hand, is where the said knowledge is processed and conclusions are drawn. In Industry 4.0, however, big data is a series of data sets that are used to draw conclusions about the goods being generated by the use of analytics. This method significantly helps to minimize decision time, improve development, increase product quality and provide machine repair/service heads-up [44]. The collection of data is achieved by sensors that are mounted inside the process at different points. These points include manufacturing equipment, processes of development, systems of company management and systems of customer management. Intelligent machinery is a necessity for a smart factory to operate continuously and provide quality goods. In addition, an important factor is the maintenance of the equipment as well. The failure of the machinery and maintenance requirements can be predicted with the assistance of the data generated by the sensors. Failure detection and routine maintenance can dramatically reduce breakdowns. Increasing the productivity of production [44].
- Robotics: Robots have been used around the world in the manufacturing industry for a long time. The primary reason for their use is the accuracy that they can work with. In addition, in contrast to humans, they can perform complex tasks in less time. Industry 4.0 supports more systems that are autonomous. Autonomous robots are thus an essential feature of I4.0 implementation in the manufacturing industry. In addition, autonomous robots are used in the production phase and are part of the smart factory. In comparison, collaborative robots work alongside humans and assist them in their work. This makes the workers more agile and increases their results [44]. The implementation of autonomous robots, however, would increase the efficiency and productivity of the process considerably. In addition to making things more viable for the entire process [50].
- Additive Manufacturing: As Industry 4.0 is the 4th industrial revolution, it brings new ways of manufactured products with it. Additive production is one of the physical components of I4.0. With the existing production practices of companies, their capacity is limited when it comes to customization and the world is moving rapidly towards customization at the moment. Additive manufacturing would also be a critical part of an industry 4.0 implementation organization [44]. An entirely new idea is additive manufacturing, and businesses are beginning to use it. Using this technology, aerospace companies are at the forefront. This method of production primarily utilizes 3D printing, the implementation of various prototypes and the production of individual units [44].
- Food Manufacturing: Within and for the survival of companies, the food processing industry has tremendous rivalry. They need to manufacture goods that are distinct from each other. This exercise pushes businesses in a certain area to come up with new products or buy a small or medium-sized company. In addition, businesses will need to adjust their processes, which

range from manufacturing to their supply chain, in order to gain a competitive advantage [51]. This includes digitizing their processes and applying lean practices within their processes to the enterprise. Large-scale businesses are still using these methods as it gives them higher efficiency, improved product quality and improved supply chains. Moreover, it is important to note that, depending on the commodity it is for, the food processing industry has various types of supply chains. There is a general supply chain, and then there is a cold supply chain. The usual supply chain should be used for all goods that are not temperature sensitive. Whereas, in the case of temperature-sensitive goods, the cold supply chain must be prevented from degrading until it hits the consumer [44].

11.11 IIoT ARCHITECTURE AND SCENARIO

It is possible to characterize the IIoT architecture in two ways, one being from the hardware and the other from the software. The hardware [1] can be divided into four main components: physical layers, networks, industrial clouds, and smart terminals.

- Physical layer: It is the most fundamental aspect of the entire system that has a direct influence on industrial output and execution. The physical layer is made up of several physical devices, such as sensors, monitors, automated guided vehicles, production equipment, and so on. The raw products are transported by AGV in compliance with their RFID to a serious destination, which shows the key parameters of the products and the data generated during the Development. To ensure the safety and integrity of the entire operation, the sensors and monitors are all responsible for documenting the product's details.
- Networks: Networks have an important role to play in converting knowledge from sensors to cloud applications to control systems. Networks are supported by many techniques, such as MCNs (mobile networking networks), industrial Ethernet, com-Wi-Fi, e.g., MOXA-Nport-W2150A, IEEE 802.11 USB-Wi-Fi module, 4G, 5G, and so on. The growth of networks speeds up the conversion time on a wide scale and provides stability.
- Industrial clouds: Industrial clouds are responsible for optimally computing the algorithm, deciding and storing the massive data produced during the computation.
- Smart Terminals: Users, owners and managers can communicate effectively, conveniently and visually with their industrial networks through the information presented by intelligent terminals (such as smartphones, LED screens, web pages). Intelligent terminals make it more effective and simpler to handle the production process.

The IIoT has three layers from the viewpoint of software: the physical layer, the control layer, and the application layer. The physical layer consists of different devices that provide machine-to-machine, human-to-machine inter-communication. The control layer between the physical layer and the application layer is like a

Using Blockchain in Resolving the Challenges Faced by IIoT 209

middleware. The main task of this layer is to collect the necessary data at the request of the application layer and, through channels, gateways, collectors, transform the data to the application layer. The application layer includes a range of developer-designed APIs to be implemented in some interesting applications. With the aid of the API, hardware-generated data can be easily exchanged so that the cost of the whole production can be minimized and the industrial system can be controlled efficiently [1]. There are several standard industry scenarios that have been introduced to boost the efficiency and productivity of the IIoT. The following ones are the three standard IIoT [1] scenarios:

- Supply chain management of manufacturing: Companies may use the technology of the IoT to obtain knowledge about the sourcing of raw materials, inventory, sales, and other information in a timely manner. It is possible for them to forecast the price trend of raw materials, supply and demand relations through Big Data analysis. In addition, it helps to strengthen and refine the supply chain management system in order to improve the supply chain's productivity and cut costs.
- Production process optimization: The pervasive sensing characteristics of the industrial Internet enhance the capacity and level of detection of production line processes, acquisition of real-time parameters, and monitoring of material use. Via data collection and processing, intelligent tracking, intelligent control, intelligent diagnosis and intelligent decision making, intelligent maintenance can be accomplished and efficiency can be improved in order to minimize energy consumption. Iron and steel firms, for example, use different sensors and communication networks to perform real-time monitoring of the width, thickness and temperature of processed goods during the manufacturing process, which helps to improve the quality of goods and optimize production processes.

Management of industrial safety production: The primary concern in modern industry is safety production. In certain hazardous production environments, such as mining machinery, oil and gas, IIoT technology can be used to mount the monitor to expand the current network monitoring platform into a systematic, transparent and diverse integrated network monitoring platform that can effectively protect the protection of industrial production

11.12 AIMS AND LIMITATIONS OF IIoT

While in many fields, the IIoT brings great benefits to the industry, it also faces many challenges in terms of safety, privacy, security, and so on. Some lawbreakers have spread their claws to the IIoT and are trying to cause economic losses in the world, by damaging it, even threatening people's lives. In 2003, one of the fearful and powerful attacks took place in the USA, which rendered the Slammer worm invalid for two vital control systems of a nuclear world. Similarly, it is difficult to imagine that if the national grid is targeted by the hacker one day, the entire city will not be able to supply electricity and it will cause so much damage and bring a lot of panic to people to render the government untrustworthy [1]. We can understand how essential security is from the above example. There are three priorities [1] setup for IIoT to ensure security and privacy:

- Availability: All the components that will be used during manufacturing, such as sensors, monitors, channels, should be accessible at all times during the production process, in order to avoid the delay that can cause incorrect intermediate measurement results or loss of productivity.
- Integrity: Integrity implies integrity of data and integrity of components. This implies the defense against tampering when someone intentionally tampers with the computing outcome that does not reach the appropriate quality level. Another is the integrity of the materials, all the materials should be involved together and function in order to guarantee the quality of goods according to the design plan in the manufacturing process.
- Intelligence: Machines should mark all the components invisibly or clearly. The mark acts like a book that tells its owner's entire life. According to these names, we can learn about the success of its operation from the past of the components. We can use these marks to track the whole manufacturing process if mistakes occur and it is easy for us to identify the position where the error is. These marks ensure protection without being physically attacked.

11.13 CHALLENGES FACED BY IIoT AND HOW BLOCKCHAIN HELPS IN RESOLVING THESE CHALLENGES

IoT is the network with a large number of computers, and the high security risks are also high. For many factors, IoT systems are higher security threats (i) these systems may not have well-defined perimeters, (ii) these systems are highly heterogeneous in terms of communication medium and protocols, and (iii) smartphone apps need installation and other user interface permissions, but these permissions may not be possible in IoT devices due to a large number of devices, etc. [35]. A significant part of maintaining trust between systems is to protect them. The approach to protecting these systems relies on assessments of danger and risk. The solutions to these threats are made up of several different forms of security systems. As there will be several different scenarios, the process of securing IoT environments is a difficult task and each scenario is composed of devices of various kinds. Each protection solution looks different from the other, as these systems can involve entities that are limited in various ways [35]. The IoT is a layered architecture that has its own security attacks on each layer. There are several security concerns and specifications that need to be addressed. Recent IoT research focuses primarily on protocols for authentication and access control, but it is important to implement new networking protocols such as IPv6 and 5G to meet potential IoT security requirements for the rapid advancement of technology.

The processing of information is the perception layer's main operation. To collect information, this layer utilizes sensors, RFIDs, barcodes, etc. Due to its wireless nature, the attacker will attack the sensor node. The key perception layer technologies are all types of sensors, such as RFID, NFC, and sensor nodes. This layer is

Using Blockchain in Resolving the Challenges Faced by IIoT   211

TABLE 11.6
Various Attacks and Their Solutions (Adapted from [35])

Attacks at Perception Layer	How To Resolve Them
Node capture	Authentication and encryptions
Malicious code Injection	Observe the actions of the running machine continuously.
False data injection	Authentication
Tampering	Prevention of physical damage of the sensors
Eavesdropping	Encryption and Access Control
Jamming	Use of low transmission power and channel surfing

divided into two parts, namely the nodes of perception (sensors, controllers, etc.) and the networks of perception interconnecting the network layer [35] (Table 11.6).

There are attacks at the Network layer [35] such as, Phishing site attack: The attacker attempts to catch the different IoT devices in this form of attack by making limited attempts. The attacker manages to steal one person's username and password, which leaves the whole IoT device vulnerable to cyberattacks. Access attack: An unauthorized individual gets access to the IoT network during this attack. The attacker will remain undetected for longer periods of time on the network. Instead of destroying the network, the purpose of this type of attack is to collect useful information. DoS attack: In this attack, the network is overwhelmed by an attacker with useless traffic, resulting in a targeted device and network resource depletion inaccessible to the user. Sybil attack: In a sybil attack, multiple identities can be created by malicious nodes in order to deceive other nodes. In this case, the attacker's aim is to take control of various areas of the network, without using any physical nodes. Hello flood attacks: A HELLO message used by a node to join a network. In order to flood the network and thereby prevent the transmission of other types of messages, the Hello Flood attack consists of forwarding a large amount of this particular message. The most popular form of attack that affects the network is also a routing attack [35].

The application is the uppermost layer that is accessible to the end user. This layer is made up of applications such as smart grid, smart city, health care system, and intelligent transport protocols. This layer has unique safety concerns that are not present in other layers, such as data theft and privacy problems. Most IoT applications often consist of sub-layers, typically referred to as an application support layer or middleware layer [35], between the network and the application layer. The attacks are summarized in Table 11.7

11.14 BLOCKCHAIN IN RESOLVING THE CHALLENGES FACED BY IIoT

The use of devices for the IoT is an important component of our modern society. IoT devices communication is secured with asymmetric key encryption that is managed by the infrastructure of the centralized certificate authority. Via a decentralized framework that questions the conventional centralized view of confidence in digital

TABLE 11.7
More Types of Attacks and Their Possible Solutions (Adapted From [35])

Type of Attacks	Methods to Resolve Them
Data theft	Data encryption, user and network authentication
Data corruption	Installation of Antivirus, firewalls, etc.
Sniffing attack	Improve Security protocols
DOS attack	Introduction of Intrusion Detection Systems and an Intrusion Protection System
Malicious code injection	Observe the actions of the running machine continuously
Reprogram attack	Protecting the programming process

systems [52], the new blockchain technology now offers a secure way to shift ownership of digital resources. Public key encryption allows encrypted communication by providing access to the intended recipient's public key. In a perfect world, with a guarantee that their keys are not tampered with, two parties will exchange their public keys directly with one another. This is not the case when, via a shared network, a new link is created. An adversary may intercept and modify non-authenticated or encrypted messages via a shared network. Since the two parties might never have interacted before, this leaves them no way to create a safe connection. Apart from transmitting keys to the devices manually, the use of a blockchain-based public key infrastructure (PKI) is a solution to these challenges [52]. The confirmation of security objectives for PKIs in the blockchain PKI demonstrates that the distribution of public keys in public blockchain ledgers is cryptographically safe. Protocols with dynamic authentication can be built using blockchain technology and distributed logging. The Cryptographic analysis provides a foundation for system security, but a systematic cryptographic analysis must be extended to ensure maximum credibility [52]. Blockchain offers ways to mitigate the challenges posed by the IoT because the perceived security concerns in the existing environment can be solved. The different functions it can serve include safe data storage, digital identification and fraud detection, as the system ensures that once the data is registered and processed, it cannot be altered or manipulated.

The need to safeguard data is one of the main IoT issues. A shared ledger as well as an unchangeable data record is generated by blockchain that helps secure and track communication or some other system-wide activity. Data security does not only extend to cryptocurrencies. It may protect patient information in other fields, such as healthcare, strengthen record keeping and, among other things, minimize the illegal delivery of prescription drugs. In addition to the risk of data loss, unauthorized changes and data theft, wherever a human element is involved in data processing and storage, there is also a probability of errors when inputting data or compiling documents. The value of blockchain is that it eliminates the need for logins and passwords. Instead, a client is issued an encoded identity, such as an SSL certificate, that can be verified on the distributed ledger. Human errors are minimized as all activities are automatically tracked and alerted to the network, including potential threats. Since blockchain is capable of monitoring and documenting communications

Using Blockchain in Resolving the Challenges Faced by IIoT 213

between all network-connected devices, it provides businesses with access to activity log data, suspicious logins, or attempted access to records that helps to tackle security challenges.

11.15 CONCLUSION

Blockchain can develop a more secure network that allows devices to connect to the Cloud or the Internet by providing solutions to bridge the gap in current IoT problems such as security, scalability, and reliability. Blockchain will take care of all the issues currently plaguing IoT solutions with cryptographic algorithms, verification of transactions and continuous threat detection and recording, and provide businesses with a more secure environment to operate without worrying about data security and privacy risks. Seeing the IoT issues facing industrial firms today can seem daunting. But overcoming these barriers offers game-changing business opportunities for these companies to improve efficiency and growth. However, with monetary investment and a commitment to developing business infrastructure and facilities aimed at full implementation, many of the difficulties of implementing IIoT can be overcome. It is essential to support infrastructure components alongside IoT technology in order to accelerate the IIoT revolution. With IIoT, more benefit can be achieved by organizations that follow a systematic approach to applying contemporary intelligent data management and analytics systems. Experts have suggested that IIoT would probably increase manufacturing rates much further with time. It is believed that IIoT participation is theoretically the driving force behind various types of revolution. Whether it is the business, end users or the workforce itself, the cost and disadvantages of both segments as part of the comprehensive automation cycle should be identical.

REFERENCES

1. Y. Shen. *The blockchain based system to guarantee the data integrity of IIoT*, Dissertation, 2018.
2. A. Pouraghily, M.N. Islam, S. Kundu, and T. Wolf. *Poster abstract: privacy in blockchain-enabled IoT devices*. In: *2018 IEEE/ACM Third International Conference on Internet-of-Things Design and Implementation (IoTDI)*, Orlando, FL, 2018, pp. 292–293.
3. L.-N. Lundbæk. *Energy-efficient and decentralized access control: a framework for embedded systems and mobility*, Thesis, 2019. Available: https://ethos.bl.uk/OrderDetails.do?uin=uk.bl.ethos.810679
4. Q. Wang, X. Zhu, Y. Ni, L. Gu, and H. Zhu. Blockchain for the IoT and industrial IoT: a review, *Internet of Things*, 10, 100081, 2020, ISSN 2542-6605, doi:10.1016/j.iot.2019.100081. Available: http://www.sciencedirect.com/science/article/pii/S2426605 1930085X.
5. A. Dorri, S.S. Kanhere, and R. Jurdak. *Blockchain in Internet of Things: Challenges and Solutions*, 2016.
6. T.M. Fernández-Caramés and P. Fraga-Lamas. A review on the use of blockchain for the Internet of Things, *IEEE Access* 6, 32979–33001, 2018.
7. N.B. Truong. *Evaluation of trust in the Internet of Things: models, mechanisms and applications*, Thesis, 2018. Available: https://ethos.bl.uk/OrderDetails.do?uin=uk.bl.ethos.755784

8. A. Sheth, P. Anantharam, and C. Henson. Physical-cyber-social computing: an early 21st century approach. *IEEE Intelligent Systems* 28, 78–82, 2013.
9. L. Atzori, A. Iera, and G. Morabito. Siot: giving a social structure to the internet of things, *IEEE Communications Letters*, 15, 1193–1195, 2011.
10. L. Atzori, A. Iera, G. Morabito, and M. Nitti, "The social internet of things (siot)–when social networks meet the internet of things: concept, architecture and network characterization," *Computer Networks* 56, 3594–3608, 2012.
11. S. Sicari, A. Rizzardi, L. A. Grieco, and A. Coen-Porisini, "Security, privacy and trust in Internet of Things: the road ahead," *Computer Networks* 76, 146–164, 2015.
12. Z. Taloyan. *Industry 4.0: value generation and adoption of digitalization and industrial IoT in production: the case of swedish production focused companies in Mälardalen*, Dissertation, 2020.
13. A. Melander and E. Halvord. *Blockchain – what it is, and a non-financial use case*, Dissertation, 2017.
14. S. Nakamoto. *Bitcoin: a peer-to-peer electronic cash system.* Available: https://bitcoin.org/bitcoin.pdf.
15. Bitcoin Magazine. *Why the bitcoin blockchain is the biggest thing since the internet.* Available: http://www.nasdaq.com/article/why-the-bitcoin-blockchain-isthe-biggest-thing-since-the-internet-cm608228.
16. T. Guo and D. Han Herzegh. *Availability of smart contracts that rely on external data*, Dissertation, 2020.
17. M. Crosby, P. Pattanayak, and S. Verma. Blockchain technology: beyond bitcoin. applied innovation review, *Applied Innovation* 2, 6–19, 2016. Available: https://j2-capital.com/wp-content/uploads/2017/11/AIR2016-Blockchain.pdf
18. D. Hongning, W. Huaimin, X. Shaoan, C. Xiangping, and Z. Zibin. *An overview of blockchain technology: architecture, consensus, and future trends.* In: *2017 IEEE 6th International Congress on Big Data*, pp. 557–564, 2017, doi:10.1109/BigDataCongress.2017.85
19. A. Joshi, and M. Han, and Y. Wang. A survey on security and privacy issues of blockchain technology. *Mathematical Foundations of Computing* 1, 121–147, 2018. doi:10.3934/mfc.2018007
20. A.M. Antonopoulos, *Mastering Bitcoin*, O'Reilly Media, 2014.
21. L. Ante, Smart Contracts on the Blockchain – A Bibliometric Analysis and Review (April 15, 2020). Available at SSRN: https://ssrn.com/abstract=3576393 or doi:10.2139/ssrn.3576393
22. F.A.K. Elghaish. *An automated IPD cost management system: BIM and blockchain based solution*, Thesis, 2020. Available: https://ethos.bl.uk/OrderDetails.do?uin=uk.bl.ethos.813732
23. J. Mason. Intelligent contracts and the construction industry. *Journal of Legal Affairs and Dispute Resolution in Engineering and Construction* 9, 04517012, 2017.
24. G.W. Peters and E. Panayi. Understanding modern banking ledgers through blockchain technologies: future of transaction processing and smart contracts on the internet of money. *Banking Beyond Banks and Money.* Springer, 2016.
25. K. Christidis and M. Devetsikiotis. Blockchains and smart contracts for the internet of things. *IEEE Access* 4, 2292–2303, 2016.
26. M. Andoni, V. Robu, D. Flynn, S. Abram, D. Geach, D. Jenkins, P. McCallum, and A. Peacock. Blockchain technology in the energy sector: a systematic review of challenges and opportunities. *Renewable and Sustainable Energy Reviews* 100, 143–174, 2019.
27. H. Watanabe, S. Fujimura, A. Nakadaira, Y. Miyazaki, A. Akutsu, and J. Kishigami. *Blockchain contract: securing a blockchain applied to smart contracts.* In: *2016 IEEE International Conference on Consumer Electronics (ICCE)*, IEEE, pp. 467–468, 2016.

Using Blockchain in Resolving the Challenges Faced by IIoT 215

28. J. Mason and H. Escott. *Smart contracts in construction: views and perceptions of stakeholders*. In: *Proceedings of FIG Conference*, Istanbul, 2018.

29. What-is-smart-contract-in-solidity?[Online]. Available: https://www.geeksforgeeks.org/what-is-smart-contract-in-solidity/?ref=lbp

30. G. Greenspan. *Smart contracts: the good, the bad and the lazy*, 2015. Available: https://www.multichain.com/blog/2015/11/smart-contracts-good-bad-lazy.

31. M. Finck, Blockchains and data protection in the European Union. *European Data Protection Law Review* 4(1), 17–35, 2018. doi:10.21552/edpl/2018/1/6

32. E. Politou, F. Casino, E. Alepis, and C. Patsakis. Blockchain mutability: challenges and proposed solutions. *IEEE Transactions on Emerging Topics in Computing* 1–13, 2019, doi:10.1109/TETC.2019.2949510

33. J. Bergquist. *Blockchain Technology and Smart Contracts: Privacy-Preserving Tools*, 2017.

34. D. Kortzon. *What are the Problems With Implementing Blockchain Technology for Decentralized IoT Authentication: A Systematic Literature Review*, 2020.

35. M. Aqeel. *Internet of Things: Systematic Literature Review of Security and Future Research*, 2020.

36. M. Burhan, R. Rehman, B. Khan, and B. Kim, IoT elements, layered architectures and security issues: a comprehensive survey. *Sensors* 18(9), 2796, 2018.

37. M. Abomhara and G. Koien. *Security and privacy in the Internet of things: current status and open issues*. In: *2014 International Conference on Privacy and Security in Mobile Systems (PRISMS)*, 2014.

38. S. Vashi, J. Ram, J. Modi, S. Verma, and C. Prakash, 2017. *Internet of Things (IoT): a vision, architectural elements, and security issues*. In: *2017 International Conference on ISMAC (IoT in Social, Mobile, Analytics and Cloud) (I-SMAC)*.

39. A. Burg, A. Chattopadhyay, and K. Lam. Wireless communication and security issues for cyber–physical systems and the internet-of-things. *Proceedings of the IEEE* 106(1), pp. 38–60, 2018.

40. X. Du, M. Guizani, K. Huang, Y. Mu, F. Rezaeibagha, X. Wang, Q. Xia, G. Yang, and X. Zhang. Building redactable consortium blockchain for industrial Internet-of-Things. *IEEE Transactions on Industrial Informatics* 15(6), 3670–3679, 2019. doi:10.1109/TII.2019.2901011

41. D. Kim. *A reverse sequence hash chain-based access control for a smart home system*. In: *2020 IEEE International Conference on Consumer Electronics (ICCE)*, 2020, doi:10.1109/ICCE46568.2020.9043090

42. R. Brandão. *A blockchain-based protocol for message exchange in a ICS network*. In: *SAC '20: Proceedings of the 35th Annual ACM Symposium on Applied Computing*, pp. 357–360, 2020, doi:10.1145/3341105.3374231

43. Z. Taloyan. *Industry 4.0: value generation and adoption of digitalization and industrial IoT in production: the case of swedish production focused companies in Mälardalen*, 2020.

44. M.S. Adil and S. Mekanic, *Industry 4.0 and the Food Manufacturing Industry: A Conceptual Framework*, Dissertation, 2020.

45. A. Sanders, C. Elangeswaran, and J. Wulfsberg. Industry 4.0 implies lean manufacturing: research activities in industry 4.0 function as enablers for lean manufacturing. *Journal of Industrial Engineering and Management* 9(3), 811–833, 2016.

46. A. Luque, M. Peralta, A. de las Heras, and A. Córdoba. State of the Industry 4.0 in the Andalusian food sector. *Procedia Manufacturing* 13, 1199–1205, 2017.

47. K. Zhou, T. Liu, and L. Zhou. *Industry 4.0: towards future industrial opportunities and challenges*. In: *12th International Conference on Fuzzy Systems and Knowledge Discovery*, IEEE, pp. 2147–2152, 2015.

48. A. Frank, L. Dalenogare, and N. Ayala. Industry 4.0 technologies: implementation patterns in manufacturing companies. *International Journal of Production Economics* 210, 15–26, 2019.
49. M. Muntjir, M. Rahul, and H. Alhumiany. An analysis of Internet of Things (IoT): novel architectures, modern applications, security aspects and future scope with latest case studies. *International Journal of Engineering Research & Technology* 6, 422–447, 2017.
50. I. Maly, D. Sedlacek, and P. Leitao. *Augmented reality experiments with industrial robot in industry 4.0 environment.* In: *2016 IEEE 14Th International Conference on Industrial Informatics (INDIN)*, pp. 176–181, 2016.
51. M. Lawrence and S. Friel. *Healthy and Sustainable Food Systems.* Taylor & Francis Group, pp. 82–92, 2020.
52. A. Sandor, *Security of dynamic authorisation for IoT through Blockchain technology,* Dissertation, 2018.

12 Internet of Things-Based Arduino Controlled On-Load Tap Changer Distribution Transformer

Krishan Arora

Lovely Professional University, Phagwara, India

CONTENTS

12.1 Introduction 217
12.2 Review of Literature 218
12.3 Block Diagram..218
12.4 Design Solutions...218
12.5 Working 220
12.6 Conclusion 222
12.7 Future Scope 222
References..222

12.1 INTRODUCTION

The Internet of Things is about connecting the unconnected things. Nowadays, all equipments running in home or in industry or in office or any place, you can imagine, need electricity. Electricity with constant voltage, current, and power in required form is given to any equipment. To avoid undesirable fluctuations, customers have to use stabilizer in homes but for each equipment one stabilizer is required which can be a difficult for being pocket friendly. Also industries too have the same concept for their equipments in their industries. Industries get their supply in three-phase form. Nowadays, the entire world is shifting toward the use of electricity rather than use of fossil fuels. In India, it is required to use a voltage of 220 V and a frequency of 50 Hz. But still we require different voltages for use in some industries as some equipments are made in other country or with different standards [1]. Whenever electrical energy is applied to a transformer, it could be proscribed correctly; therefore, to keep the constancy of the voltage supply on the basis of the capability of the transformer, the author utilizes the tapping notion. Tap changers can be associated at different junctions in a transformer to either primary or secondary windings. It becomes easier to access high-voltage windings when a tap is placed on the HV side because HV is offended with LV and also there is reduction in danger of lightening when breaking down. So to break this barrier we are making on-load tap changing transformer which can be controlled by Arduino controller [2].

DOI: 10.1201/9781003102267-12

12.2 REVIEW OF LITERATURE

Transformers are indispensable for generation, transmission, and distribution of alternating power. Technically, power transformer is a combination of two or more windings; through electromagnetic induction, it transmits the voltage and current to another system to serve the purpose of transmitting power. Refashioning in transformation ratio helps in regulation of voltage at tap windings, so revising in quantity of turns. The procedure of changing the transformation ratio with the help of tappings at the windings is entitled as tap changing, so that amount of turns can vary. The progression of varying the ratio of transformation by tapping the windings is known as tap changing. The method of offload tap changing is not suitable for large power supply systems. This system calls for power transformer with a voltage regulating winding, the tapping of which is changed over under loaded condition by on-load tap changer. Schemes employed for on-load tap changer involved the use of more complicated and expensive tap changing equipment [3].

12.3 BLOCK DIAGRAM

In this block diagram, the basic circuit operation is represented [4]. As shown in Figure 12.1, authors have used three-phase supply which is given to system and then phase selector to selects one phase as per requirement of output than stepping up or stepping down is done if required and then output is given. All this operation is controlled by the controller.

12.4 DESIGN SOLUTIONS

In the early stages of the research work, various design solutions were employed [5]. The initial design solution can be seen in Figure 12.2.

The tap changing transformer can do step up or step down if they have tapping in secondary windings. In the above design shown firstly it is asked what output voltage is required than tap is changed of the transformer by controller keeping in mind what voltage is given in input. Limitation of this design is that it can only make variation in output voltage as per the tapings made in the transformer which are in secondary windings. In this case, output voltage is not independent of input voltage. So author

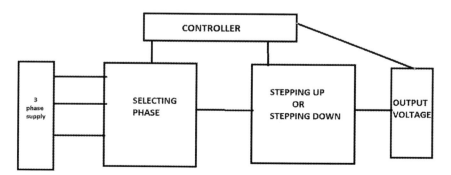

FIGURE 12.1 Block diagram of phase changing and tap changing.

IoT-Based Arduino Controlled On-Load Tap

FIGURE 12.2 Design solution for tap changing transformer.

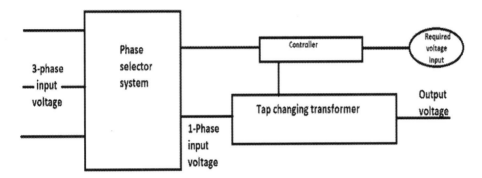

FIGURE 12.3 Design solution for tap changing transformer with phase selector system.

may require more number of voltages on input side. Next design was varied to suit these needs, as shown in Figure 12.3. The taps are lead or associates provided at different junctions on the winding. The turns ratio diverge from one tap to another and consequently dissimilar voltages can be obtained at every tap [6].

In this above design tap changing transformer system is added with a phase changing transformer to have variations in input voltage which will help in have more precise and accurate output. This system too have a limitation after having phase selector system that one-phase input voltage can step up or step down depending on which transformer is used. In case the author uses step-up transformer but as per output required author needs to do step down, then this system will fail. So to tackle this limitation next design was made [7], as shown in Figure 12.4.

In the design shown in Figure 12.4, the author can step up and step down during load conditions depending on the voltage required for the output but precision of output voltage would be compromised. In case the author needs output of 220 V but all the voltages in three-phase system are 60, 150, and 250 V, then author can step down 250 V to 220 V but it would require a lot of tapings in secondary windings which could create a problem to control it through a microcontroller. So to overcome this problem, author has replaced combination of step up and step down transformer with an autotransformer that can act as both step up and step down by changing the direction of input and output. The final design was developed [8], as shown in Figure 12.5.

This final design uses a servo motor that rotates the autotransformer to the desirable output voltage. Also, combination of relays are used to change the input and the

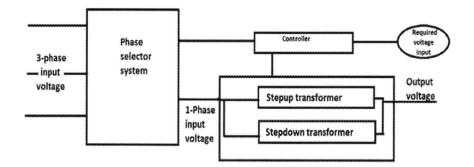

FIGURE 12.4 Design solution for tap changing combination transformer with phase selector system.

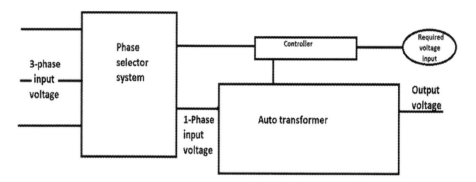

FIGURE 12.5 Design solution for tap changing autotransformer with phase selector system.

output side of the autotransformer to make it step up autotransformer or to step down autotransformer as per required of the system.

12.5 WORKING

This research work focuses on the planning and construction of Internet of Thing based phase changing and automatic on-load tap changing distribution transformer. This ensures a continuous offer of power to the load as per the input provided until it's asked to prevent it. In the next step needed voltage is compared with incoming voltage than Arduino mega decides; however, the connected autotransformer would work, i.e., it ought to work as step up transformer or step down transformer [9]. As Arduino mega decides however transformer would work and switch the relays on and off as per demand. Additionally, voltage sensing element connected to output offers the feedback to Arduino mega that what quantity output voltage is returning. As per the system output voltage ought to be unbroken constant. Thus, the equipment connected to output will work perpetually without any drawback [10].

In this circuit, three different autotransformer acts as three phases that are received by the industry for industrial purpose as shown in Figure 12.6. Also, each

IoT-Based Arduino Controlled On-Load Tap

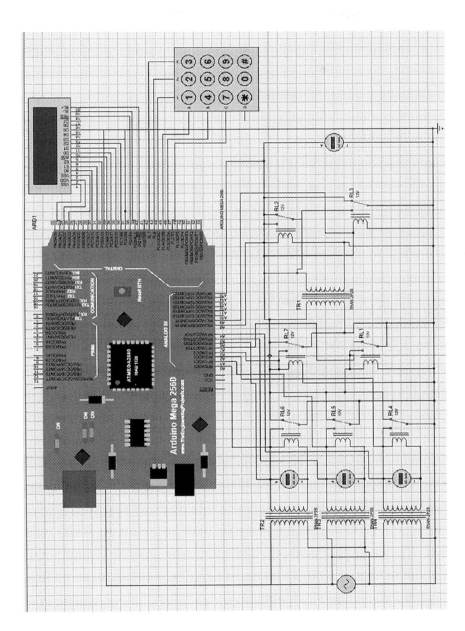

FIGURE 12.6 Circuit representation of on-load taps changer.

autotransformer is attached to AC voltage sensor in parallel [11]. Also, each AC supply coming from autotransformer is connected with the relay module which acts as a switch and for our phase selector circuit. All the three relays are controlled by the Arduino mega which acts as its brain and response as per required voltage. In the next part of this circuit where author gets a single-phase supply, it can be step up or step down the voltage of the single-phase supply as per the requirement of the output voltage. In this circuit, an error sensing feedback control called servo is used to correct system performance [12]. This typically requires sophisticated controllers, often a dedicated module specifically designed for use with servomotors.

12.6 CONCLUSION

This research focuses on the design and construction of Internet of Things based phase changing and automatic on-load tap changing distribution transformer. This ensures a continuous supply of power to the load as per the input provided till it is asked to stop it. The supply coming to industry in form of three phases in the industry [13]. So this system firstly asks for input voltage which is required to operate the equipment after the required voltage is input, then the system compares the required voltage with all the three voltages in three-phase system and whichever voltage in three-phase system is nearest to the required voltages that are forwarded to next step [14]. In the next step required voltage is compared with incoming voltage, then Arduino mega decides how the connected autotransformer would work, i.e., whether it should work as step-up transformer or step-down transformer. As Arduino mega decides how transformer would work and switch the relays on and off as per requirement. Also voltage sensor connected to output gives the feedback to Arduino mega that how much output voltage is coming. As per our system output voltage should be kept constant. So our equipment connected to output can work constantly without any problem.

12.7 FUTURE SCOPE

> This research can be used in any type of industry as well as in homes too.
> As there is a trend of electric vehicles in the present scenario than our system can be used in electric vehicles for charging them as they different electric vehicles would require different voltages to charge them.
> This system can be used as an external device for any equipment as well as it can be installed inside in any equipment too.

REFERENCES

1. G.R.C. Mouli, P. Bauer, T. Wijekoon, A. Panosyan, "Design of a Triac OLTC for grid voltage regulation". *IEEE*, 2(6), 2277–7970, 2012.
2. S.V.M. Bhuvanaika Rao, B. Subramanyeswar, "Fine voltage control using OLTC by static tap change mechanism". *International Journal of Advance of Computer Research*, 2, 328, 2002.

IoT-Based Arduino Controlled On-Load Tap

3. S. Vigneshwaran, T. Yuvaraja, "Voltage regulation by solid state tap change mechanism for distributing transformer". *International Journal of Engineering Research & Technology (IJERT)*, 4(02), 2015.

4. C.V. Thomas, P.M. Noufal, P.N. Mohan, B.S. Vinod, K.A. Eldhose, "Solid state on load tap changer for transformer using Arduino". *International Research Journal of Engineering and Technology (IRJET)*, 04, 2017.

5. S.M. Bashi, "Microcontroller-based fast on-load semiconductor tap changer for small power transformer". *Journal of Applied Sciences* 5(6), 999–1003, 2005.

6. A. Stigant and A.C. Franklin, *J & P transformer book*, 12th ed., Elsevier, Butterworth.

7. Maschinenfabrik Reinhausen Gmb H, On-Load Tap-Changers for Power Transformers – A Technical Digest, 2009.

8. J.H. Harlow, *Electric power transformer engineering*, 2nd ed., Taylor & Francis Group, 2006.

9. J. Faiz and B. Siahkolah, *Electronic tap-changer for distribution transformers*, Springer, 2011.

10. P. Bauer and S.W.H. De Haan, "*Solid state tap changers for utility transformers*". In: *IEEE AFRICON 1999*, vol. 2, pp. 897–902, 1999.

11. P. Bauer and S.W.H. De Haan, "*Electronic tap changer for 500 kVA/10 kV distribution transformers: design, experimental results and impact in distribution networks*". In: *The 1998 IEEE Industry Applications Conference*, vol. 2, pp. 1530–1537, 1998.

12. P. Bauer and R. Schoevaars, "*Bidirectional switch for a solid state tap changer*". In: *2003 IEEE 34th Annual Power Electronics Specialist Conference (PESC)*, vol. 1, pp. 466, 471, 2003.

13. D. Gao, Q. Lu and J. A. Luo, "*New scheme for on-load tap-changer of transformers*". In: *International Conference on Power System Technology*, pp. 1016–1020, 2002.

14. G.R. Chandra Mouli, P. Bauer, T. Wijekoon, A. Panosyan and E. Bärthlein, "Design of a power electronic assisted OLTC for grid voltage regulation". *IEEE Transactions on Power Delivery*.

Index

Page numbers in **bold** indicate tables, page numbers in *italic* indicate figures.

A

abnormalities, 41, 83, 84
access to the cloud, 24, 30, 36, 67, 136, 165
accuracy, 2, 9, 27, 85, 103, 144, 150, 152, 153, 175, 176, 178, 180–185, 201, 207
actuators, 5, 20, 24, 34, 50, 58, 111, 169, 189, 190, 199, **205**
adjustable tapping, 217, 218
advancement in microchip technologies, 161
agricultural industry, 37, 143–158
agricultural problems, 150–154
agriculture, 21, 25–28, 57, 129, 144, 146, 148, 150–155, 157, 158, 162
alarm, 59, 64, 66, 72, 78, 80–84, 170, 200
android, 56, **132**, 134, 135, 181
appearance variations, 81, 82, 124, 146, 152, 179, 218, 219
application of Internet of Thing (IoT), 1, 2, 57–59, 88, 104, 150–154, 158, 167, 168, 172, 189–191, 198, 200–202, 204, 206, 211
Arduino controller, 141, 217–222
artificial intelligence (AI), 20, 26, 31, 37, 144, 145, *145*, 148, 150, 154, 158, 163, 167, 169, 171, 176, 182, **184**
attacks, 36, 89, 101, 102, 104, 154, 189, 198, 202, 209–211, **211, 212**
authentication, 36, 89, 92, 101, 103, 135, 141, 154, 198, 202, 203, 210, **211**, 212, **212**
automation, 20–22, 26, 31, 33, 57, 59, 66, 108, 110, 113, 116, 117, 125, 129–141, 146, 152, 162, 163, 176, 198, 200, 213
   for the consumer, 162, 198
auto power cut-off, 139, 163, 220, 222

B

bandwidth management, 10, 27, 47, 91, 99, 100, 131, 166, 168, 169
big data, 20, 22, 28, 31, 37, 67, 109, 111–113, 115, 117, 162, 190, 201, **205**, 206, 207, 209
blockchain, 36, 37, 189–213
border supply chain, 162
business decisions, 20
business processes, 161–163, 172

C

capital, 24, 29, 30, 109, 192
change in routing, 102

circuit, 63, 220, *221*, 222
cloud computing, 5, 21, 30, 57, 111, 117, 136, *164*, 165–167, 173, 198, 199, **205**, 206
cloud server, 165, 166
cognitive-IIoT, 168–173
comfort, 130, 131, 190, 198–200
computational power, 117, 202
computer-decoded text, 175
construction, 71, 161, 200, 220, 222
continuous growth, 57
control, 3, 4, 7–10, 12–14, 20, 22, 23, 27, 28, 33–35, 42, 48, *50*, 51, 56, 58, 59, 61, 63, 65, 66, 84, 92, *94*, 97–99, 111, 113, 115, **121**, 122, 130, **132**, 133, 136–138, *140*, 141, 150–152, 155, 156, 165, 167, 190, 196, 200–204, **205**, 208–210, **211**, 219, 222
   capabilities, 169
controlling industrial operations, 7, 20
controlling of industrial processes, 1, 34
controlling of voltage, 217–219, 222
cost, 1, 3, 9, 11, 14, 20–24, 26–28, 31, 32, 35, 43, 50, 51, 56, 60, 61, 66, 67, 85, 88, **98**, 108, 109, 113, **114**, 116, 118–121, **121**, 123, 124, 130, 131, **132**, 134, 145, 146, 150–152, 156–158, 166, 168, 169, 171, 190, 204, 209, 213
CR-LPWAN, 171, 172
CRN in IIoT, 95, 169
cyberattacks, 33, 211
cyber-physical systems (CPS), 21, 108, 110, 111, *112*, 162, **205**

D

data integrity, 34, 204
data ownership, 212
decision-making, 20, 22, 27, 34, 35, 108, 110, 111, 118, 146, 148, 157, 162, 191, 203, 204, **205**
deep learning algorithms, 31, 111, 144, 148, 150, 180, 181
degree of privacy, 198
DHT11 temperature sensor, 56, 59, 61, 62, 67, 68, **68**, 69, *69*, 82, 135
digital convergence, 162
digitization impact, 24, 107, 111, 124
distant locations, 95, 96, *96*
distribution, 111, 122, 161, 179, **205**, 212, 217–222
   transformer, 217–222
dynamic IP allocation, 42

225

226 Index

E

ecosystem, 108, 190, 191, 203
edge computing, *164*, 165–167
edge network, 161–173
efficient management, 1, 157
electrical power, 135, 217
end-to-end reachability, 12, 24, 34, 35, 115, 190, 191, 203
energy generation, 161
ESP 8266, 57, 59, 60, *61*, 64, 77

F

feature extraction, 179, 180, 182–184
firebase, 134
fog computing, *164*, 165–167, 173
fragmented delivery of knowledge, 34

G

goods, 29, 34, 36, 112, 113, 121, 124, 162, 200, 203, 207–210

H

handwriting recognition, 175, 176
handwritten text converter, 175
healthcare, 9, 10, 21, 30, 31, 36, 37, 41–52, 212
high level of digital intelligence, 161
home appliances, 59, 129–131, 133–135, 199
home automation system (HAS), 59, 129–141

I

ICT, 107, 115
IIoT architecture, 5–9, 15, 163, *164*, *172*, 208, 209
IIoT computing strategies, 165, 166
IIoT devices, 27, 29, 33, 36, 162, 165–167, 169, 171
IIoT edge devices, 163, *164*, 166, 167
IIoT platforms, 163, 164
improved productivity, 21, 22, 28, 108, 113, **114**, 115, 125, 144, 176, 203, 208, 209
independent robots, 155
industrial IoT, 1–15, 19–37, 108, 116, 144, 161–173, 176, 177, *177*, 183, 189–213
industrial revolution, 4, 7, 20, 22, 32, 108, 111, 115, 116, 162, 163, 191, 204, 207
industries, 20–22, 33–35, 37, 56, 59, 108, 109, 122, 144, 161–163, 165, 167, 169, 189–191, 203, 204, 217
businesses, 107
Industry 4.0, 2–5, 7, 8, 20–22, 37, 51, 52, 107–125, 161–163, 191, 203, 204, **205**, 206, 207

intelligent network of equipment, 162, 189, 204, 206, 207
intensive industries, 108, 161, 204
internet, 2, *4*, 5, 8, 20, 24, 25, 30, 32, 33, 37, 41, 48, 57–60, 67, 75, 76, 87–104, 107, 109, 111, 130, 148, 152, 154, 165, 171, 176, 180, 183, 189, 192, 198, 199, 201, 205, 206, 209, 213, 220
consumer, 24, 57, 162, 198
internet-connected smart device, 20, 32, 37, 189, 199, 200, **200**, 205
IoT technologies, 47, 57, 67, 152, 154, **200**, 209, 213

J

Java programming language, 134

L

lack of expertise, 157
lean management, 118, **121**, 122
lean manufacturing, 107–125
load balancing, 52
logistics, 20, 24, 28, 29, 35, 113, 116, 162, 206
industries, 162
LPWAN for IIoT, 171, 172

M

machine learning (ML), 20, 31, 34, 111, 144, 162, *164*, 182, 183
machine vision technology, 143–158
machinery, 3, 7, 20, 115, 144, 161, 166, 204, 207, 209
maintenance fees product miniaturization, 22, 107, 113
manufacturing cycle, 2, 3, 110, **114**, 121, 125, 213
manufacturing processes, 21, 118, 120, 123–125, 161, 163, 168, 209, 210
microcontroller, 57, 60, 63, 131, 135, 219
mining, 10, 110, 161, 192, 209
mobile, 20, 34–36, 58, 88, 90, 103, 113, 130, **132**, 133–135, 137, 148, 163, 167, 190, 198, 199, 206
monitoring, 1, 3, 7, 20, 22, 26–28, 30, 34, 49, 55–85, 92, 94, 111, 115, 125, 134, 145, 151–153, 156, 162, 163, *164*, 167, 195, **205**, 209, 212
MQ-2 gas sensor, 56, 70, 71, 77–79, 81, 82
multilingual scripts, 183, 185

N

network function virtualization (NFV), 41, 43
neural network, 148, *149*, 150, 176, 180–183

Index   227

O

offline handwritten character recognition techniques, 175, 179
on-load tap changers, 217–222
operational technology (OT), 20, 163
operational time, 1
operations reliability, 20, 21, 60, 113, 146, 190, 191, 203, 213
optical character recognition (OCR), 146, 175–185
optimizing processes, 20, 104, 108, 109, 115, 118, 125, 179
output voltage, 71, 218, *219*, 220, 222

P

people, 2, 30, 34, 57, 58, 71, 72, 78, 79, 88, 108, 111, 120, 124, 161, 162, 165, 183, 191, 198–201, 205, 209
PIR motion sensor, 56, 59, 62, 71–75, 79, 81, 108, 109, 148, 202
power consumption units, 32, 33, 63, 69, 103, 104, 139, 168
powerful network capabilities, 33, 149
preciseness, 144
precision in OCR, 175–185
product-driven approach to production, 107
productivity, 1, 2, 9, 20, 21, 26, 27, 29, 50, 58, 108, 113, 123, 125, 144, 146, 151, 161, 162, 176, 198, 203, 207, 209, 210
products, 29, 107, 109, 110, 112–115, **121**, 122, 124, 144, 152, 162, 166, 199, 200, 207, 208

R

radio-frequency identification (RFID), 28, 29, 31, 116, 117, 162, 163, 190, 198, 199, **200**, 208, 210
realization of digital transformation, 22, 162, 203
real-life application, 175, 176
real-life problems, 176
real-time database, 135, 136, 141
real-time monitoring, 1, 20, 209
recognition efficiency, 176, 178
recovery, 28, 198
revolutionary applications, 4, 7, 204

S

safety in smart manufacturing, 161
SDN-based IoT devices, 42
security, 3, 5, 9–12, 22, 24, 25, 28, 31, 33–37, 41, 49–51, 59, 61, 65, 72, 76, 85, 89, 101, 103, 104, 113, 130, 131, 148, 154, 163,

166, 169, 190, 191, **198**, 199–202, **205**, 206, 209, 210, 212, 213
sensor devices, 5, 7, 9, 10, 14, 20, 22, 24–31
services, 15, 28, 41, 50, 58, 67, 85, 88, 89, 92, 94, 109, 117, 136, 152, 161, 165, 167, 171, 190, 191, 198. 200, 203, 204, **205**
sharing information, 14, 36, 167, 169
single-phase transformer, 222
smart farming, 144, *145*, 146, 150, 151, 153, 154, 158
Smart Home, 129–141, 162, 200, 206
smart manufacturing, 111, 161, 162, 168
smart objects, 1, 7, 107
software-defined network (SDN), 41, 42
spectrum allocation issue, 169
spectrum scarcity issue, 161–173
supply chain, 20, 21, 23, 24, 28–30, 36, 37, 109, 110, 123, 125, 162, 208, 209
switching procedure, 169
synergies, 2
system integration, 20

T

tap changer, 217–222
telecommunications, 161, 199
threshold consumption units, 23, 170, 179
tiny and cheap devices, 161
tracking people, 27, 29, 31, 94, **99**, 115, 116, 146, 152, 162, 163, 209
traditional farming methodologies, 25

U

unprecedented attention, 35, 117

V

variant of cognitive-IIoT, 168–171, *172*, 173
violation, **99**

W

waste, 9, 27, 57, 108, 109, **114**, 118–120, **121**, 122, 124, 152, 161, 197
    reduction, 118, 119
water management, 57, 145, 161
Wi-Fi module, 75, 135, *135*, 208
wireless, 9, 25, 31–36, 56, 66, 88, 89, 91–93, 97, **98, 99**, 111, 113, 116, 130, 131, 133, 157, 161–163, 168, 169, 189, 200, 201, 206, 210
    networks, 25, 31, 33, 36, 111, 161, 162, 168, 206

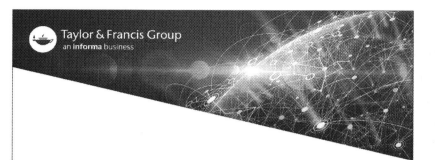

Taylor & Francis eBooks

www.taylorfrancis.com

A single destination for eBooks from Taylor & Francis with increased functionality and an improved user experience to meet the needs of our customers.

90,000+ eBooks of award-winning academic content in Humanities, Social Science, Science, Technology, Engineering, and Medical written by a global network of editors and authors.

TAYLOR & FRANCIS EBOOKS OFFERS:

- A streamlined experience for our library customers
- A single point of discovery for all of our eBook content
- Improved search and discovery of content at both book and chapter level

REQUEST A FREE TRIAL
support@taylorfrancis.com

Printed in the United States
by Baker & Taylor Publisher Services